Energy Harvesting

A thorough treatment of energy harvesting technologies, highlighting radio frequency (RF) and hybrid-multiple technology harvesting. The authors explain the principles of solar, thermal, kinetic, and electromagnetic energy harvesting, address design challenges, and describe applications. The volume features an introduction to switched mode power converters and energy storage and summarizes the challenges of different system implementations, from wireless transceivers to backscatter communication systems and ambient backscattering. This practical resource is essential for researchers and graduate students in the field of communications and sensor technology, in addition to practitioners working in these fields.

Apostolos Georgiadis is Honorary Associate Professor at Heriot-Watt University, Edinburgh, UK. He is a former editor-in-chief of *Wireless Power Transfer* (Cambridge). He is an EU Marie Curie Fellow, an URSI Fellow, chair of URSI Commission D Electronics and Photonics, and a distinguished lecturer of the IEEE Council on RFID.

Ana Collado was Assistant Professor at Heriot-Watt University, Edinburgh, UK. She is an EU Marie Curie Fellow.

Manos M. Tentzeris is Ken Byers Professor in Flexible Electronics with the School of Electrical and Computer Engineering (ECE), Georgia Tech. He is a Fellow of IEEE, a Fellow of the Electromagnetics Academy, and a member of URSI Commission D Electronics and Photonics and of the Technical Chamber of Greece.

EuMA High Frequency Technologies Series

Series Editor
Peter Russer, Technical University of Munich

Homayoun Nikookar, *Wavelet Radio*
Thomas Zwick, Werner Wiesbeck, Jens Timmermann, and Grzegorz Adamiuk (Eds), *Ultra-wideband RF System Engineering*
Er-Ping Li and Hong-Son Chu, *Plasmonic Nanoelectronics and Sensing*
Luca Roselli (Ed.), *Green RFID Systems*
Vesna Crnojević-Bengin (Ed.), *Advances in Multi-band Microstrip Filters*
Natalia Nikolova, *Introduction to Microwave Imaging*
Karl F. Warnick, Rob Maaskant, Marianna V. Ivashina, David B. Davidson, and Brian D. Jeffs, *Phased Arrays for Radio Astronomy, Remote Sensing, and Satellite Communications*
Philippe Ferrari, Rolf Jakoby, Onur Hamza Karabey, Gustavo Rehder, and Holger Maune (Eds), *Reconfigurable Circuits and Technologies for Smart Millimeter-Wave Systems*
Apostolos Georgiadis, Ana Collado, and Manos M. Tentzeris, *Energy Harvesting*

Energy Harvesting

Technologies, Systems, and Challenges

APOSTOLOS GEORGIADIS
Heriot-Watt University, Edinburgh

ANA COLLADO
Formerly Heriot-Watt University, Edinburgh

MANOS M. TENTZERIS
Georgia Institute of Technology

CAMBRIDGE
UNIVERSITY PRESS

CAMBRIDGE
UNIVERSITY PRESS

University Printing House, Cambridge CB2 8BS, United Kingdom

One Liberty Plaza, 20th Floor, New York, NY 10006, USA

477 Williamstown Road, Port Melbourne, VIC 3207, Australia

314–321, 3rd Floor, Plot 3, Splendor Forum, Jasola District Centre, New Delhi – 110025, India

79 Anson Road, #06–04/06, Singapore 079906

Cambridge University Press is part of the University of Cambridge.

It furthers the University's mission by disseminating knowledge in the pursuit of education, learning, and research at the highest international levels of excellence.

www.cambridge.org
Information on this title: www.cambridge.org/9781107039377
DOI: 10.1017/9781139600255

First published 2021

Printed in the United Kingdom by TJ Books Limited, Padstow Cornwall

A catalogue record for this publication is available from the British Library.

Library of Congress Cataloging-in-Publication Data
Names: Georgiadis, Apostolos, author. | Collado, Ana, author. |
Tentzeris, Manos M., author.
Title: Energy harvesting : technologies, systems, and challenges /
Apostolos Georgiadis, Ana Collado and Emmanouil M. Tentzeris.
Description: First edition. | Cambridge ; New York, NY : Cambridge
University Press, [2020] | Series: EUMA high frequency technologies
series | Includes bibliographical references and index.
Identifiers: LCCN 2020023799 (print) | LCCN 2020023800 (ebook) |
ISBN 9781107039377 (hardback) | ISBN 9781139600255 (epub)
Subjects: LCSH: Energy harvesting.
Classification: LCC TK2896 .G46 2020 (print) | LCC TK2896 (ebook) |
DDC 621.042–dc23
LC record available at https://lccn.loc.gov/2020023799
LC ebook record available at https://lccn.loc.gov/2020023800

ISBN 978-1-107-03937-7 Hardback

To our daughters Ariadne, Markella-Renata, and Christina-Apostolia

Contents

Preface

This book discusses energy harvesting technologies for low-power wireless sensing platforms. This is a multidisciplinary topic requiring background from different physics-related disciplines such as thermodynamics and mechanics but also electrical engineering, optimization, and signal processing. In addition, material science and additive manufacturing has further helped develop low-cost sensors and circuits that enable form and cost reduction of such platforms and their ubiquitous application.

Our involvement in the field began around 2009. Apostolos and Ana had already been collaborating since 2004 in Spain in the context of a different research field, and around 2009 we decided to begin working on radio frequency (RF) energy harvesting. At this time, we were introduced to Manos during a meeting of the European Union (EU) European Cooperation in Science and Technology (COST) IC0803 project that we were running. From this moment, our collaboration in the field began, and over the last ten years we have produced numerous publications related to different aspects of energy harvesting.

Chapter 1 is an introductory chapter providing a brief overview and a perspective to the research and industrial possibilities related to such a multidisciplinary field of energy harvesting. Chapter 2 is devoted to 2D-3D integration of energy autonomous sensors using inkjet printing fabrication. Chapters 3–5 discuss solar, kinetic and thermal energy harvesting. Chapters 6 and 7 are devoted to wireless power transmission and RF energy harvesting. Chapter 8 discusses dc voltage conversion and power storage, and finally, Chapter 9 is devoted to a system overview of wireless sensing platforms with energy harvesting capability. As part of a course, one could begin with Chapter 1, follow with Chapter 2, and then cover the chapters related to the different energy harvesting technologies. Chapters 3 through 5 and the combined set of Chapters 6 and 7 could be taught in any order. Chapter 2 may also be offered after Chapters 3 through 7. One could then cover the final Chapters 8 and 9.

We would like to thank our numerous former students who have contributed to our work on energy harvesting. First and foremost, Sangkil Kim, now assistant professor at Pusan National University, South Korea, has contributed to a great number of publications and provided the main draft of Chapter 2 based on his expertise in inkjet printing. Special thanks go to Dr. Spyros Daskalakis, Heriot-Watt University, for his help with the book editing. Our thanks also go to our former students, collaborating students and visiting students, Francesco Giuppi, Huawei, Milan; Kyriaki Niotaki, Maynooth University; Gianfranco Andia Vera, Multiwave; Rushi Vyas, University of Calgary; John Kimionis, Nokia Bell Labs, New Jersey; Alirio Boaventura, Daniel Belo, Ludimar Guenda and Ricardo Fernandes from the University of Aveiro; Marco Virili, Qorvo, Inc.; Valentina Palazzi, University of Perugia; Massimo del Prete, University of Bologna; Chiara Mariotti, Infineon Technologies; Maria Valeria de Paolis, Laboratory for Analysis and Architecture of Systems (LAAS) of the French National Center for Scientific Research (CNRS); Martin Schuetz, University of Erlangen–Nuremberg; Ferran Bolos, Javier Blanco, Angel Servent, Ernest Silvestre, Ricard Martinez, Omar Andre Campana Escala, Cesar Meneses Ghiglino, and Gustavo Adolfo Sotelo Bazan from the Polytechnic University of Catalunya, Castelldefels; and Sergi Rima from Rovira i Virgili University. We apologize for potential errors in affiliations; there are just too many of you to accurately keep track!

We would also like to thank our numerous collaborators: Professor Nuno Borges Carvalho, Aveiro University; Professor Luca Roselli, University of Perugia; Professor Yoshihiro Kawahara, University of Tokyo; Professor Alessandra Constanzo, University of Bologna; Professors Lauri Sydanheimo, Leena Ukkonen, and Toni Bjorninen, Tampere University; Professor Zoya Popovic, University of Colorado, Boulder; Professor John Sahalos, University of Nicosia; Professor Smail Tedjini, Grenoble Institute of Technology; Professor Hendrik Rogier, Ghent University; Professors Luciano Tarriconne and Giuseppina Monti, University of Salento; Professors Antonio Lazaro and David Girbau from Rovira i Virgili University; and Professor George Goussetis, Heriot-Watt University. We for your collaboration and fruitful discussions throughout this period but foremost for your friendship.

Finally, we would like to express our gratitude to Julie Lancashire from Cambridge University Press for her continuous support and patience as well as to Sarah Strange and the rest of the editorial staff at Cambridge University Press.

We hope that the book can provide a starting point for our readers to the world of energy harvesting, and we welcome any comments and feedback!

1 Introduction

1.1 Wireless Sensing Platforms

Energy harvesting technologies have spurred interest in the academic and industrial communities due to the emerging applications of miniature, low-power, wireless sensors. Concepts such as the Internet of Things (IoT) envision myriad networked devices used to integrate and automate homes, offices, factories – in other words, everything [1, 2]. One of the challenges for such devices is their ability to operate for long periods of time autonomously, i.e., without the need to be connected to a wired power supply or to substitute or recharge their batteries.

A notable milestone toward miniature wireless sensor nodes has been the smart dust project by University of California, Berkeley, researchers in the late 1990s [3]. The smart dust project introduced the concept of autonomous sensing and communication cubic-millimeter-sized motes (i.e., small particles) forming a massive distributed sensor network [3]. As a result, several wireless sensing platforms have been developed in an attempt to implement the smart dust concept. Widely popular implementations of such sensing platforms, albeit without achieving the ultimate cubic millimeter volume vision, have been the Mica mote [4] and subsequently the Telos mote [5] integrating a low-power microcontroller, sensor interface circuitry and a radio transceiver.

An alternative technology toward implementing ultralow-power wireless sensing platforms has been radio frequency identification (RFID) technology based on radar principles and backscatter communication [6]. In such systems, passive sensing and identification tags comprise ultralow-power radio transceivers that operate based on antenna load modulation that does not require an amplifier because the necessary power for both powering the tag and for communication is provided by an interrogator reader device [6].

Finally, the increased interest for ultralow-power, energy autonomous, wireless sensing platforms has been further fueled by the fifth generation (5G) communication systems that attempt to implement a massive number of interconnected devices communicating at low bit rates [7]. As the number of interconnected devices and potential for new applications keeps increasing, it is only natural to expect that the interest for energy autonomous, wireless sensing platforms will continue into the sixth generation (6G) systems and beyond.

Figure 1.1 Nikola Tesla, with his equipment for producing high-frequency alternating currents. Credit: Wellcome Collection. Attribution 4.0 International (CC BY 4.0)

1.2 Energy Harvesting Revolution

The physics behind the commonly employed energy harvesting technologies, solar, mechanical, thermal, and radio frequency, has been known for many years. For example, more than 100 years ago, Tesla envisioned the wireless transmission of power. Figure 1.1 shows a celebrated photo of Tesla in his laboratory in Colorado Springs. In another example, more than 100 years ago, Gulielmo Marconi began his experiments in wireless telegraphy in Vila Griffone, Bologna, Italy. The photo shown in Figure 1.2 is a setup of his laboratory in the Marconi Museum, where one can see on his desk a disc-shaped device that was a thermocouple, a thermoelectric generator that he used in his experiments.

One could probably come up with numerous other examples. Advances in materials and fabrication techniques have enabled the miniaturization and the performance improvement of such energy harvesting devices that make them suitable for low-power wireless sensors. Combined with advances in electronic design and integrated circuit technologies that have led to the reduction of operating power of electronic circuits, the vision of energy harvesting powered wireless sensor platforms becomes more and more possible.

1.3 This Book

The topic of energy harvesting technologies is very broad and diverse, given that each of the energy harvesting technologies represents a completely different field

Figure 1.2 Marconi's laboratoy at Villa Griffone, Bologna, Italy.

of research. This book discusses the main energy harvesting technologies, namely solar, kinetic, thermal, and electromagnetic (EM), together with an introduction to power converters and energy storage. We try to provide an answer to questions such as how much power can be harvested and what are the main challenges in implementing these harvesting systems.

Table 1.1 presents indicative performance results from different types of energy harvesters with emphasis on low-profile transducers suitable for micropower generation. There exists a large variation among the size of the transducers and the amount of energy that can be generated. As a result, the final selection of the employed type of transducer depends greatly on the application requirements and scenario, which makes the presented results of Table 1.1 only indicative of the potential of the various harvesting methods.

Table 1.1 Indicative harvested power values from different transducer types [8].

Energy source	Harvested power	Conditions / available power
Light / solar	60 mW	6.3 cm × 3.8 cm flexible a-Si solar cell AM1.5 Sunlight (100 mW/cm^2) [9]
Kinetic	8.4 mW	Shoe-mounted piezoelectric [10]
Thermal	0.52 mW	Thermoelectric generator (TEG), $\Delta T = 5.6$ K [11]
Electromagnetic	1.5 μW	Ambient power density 0.15 μW/cm^2 [12]

Each transducer technology has distinct advantages and disadvantages. For example, solar energy is ubiquitous, whereas solar harvesting is challenging in indoor scenarios and during night or cloudy conditions. Thermal energy harvesters are typically hampered by a low transducer efficiency, especially when a low-temperature gradient is present, while kinetic energy harvesters are sensitive to the natural vibration frequencies of the harvester and application settings.

When it comes to ambient EM energy harvesting, the available energy density is usually orders of magnitude below the corresponding values of the other energy sources, although measurement campaigns in crowded urban settings have shown the possibility of harvesting a useful amount of EM energy from the ambient [13, 14, 15], especially using wideband or multiband harvesters. Nonetheless, EM energy harvesters are intimately related to systems exploring intentional EM radiation to power up electronic devices, wireless power transfer, with RFID technology being a notable application example that already enjoys commercial success.

The dc voltage output of the various energy harvesting transducers can vary significantly from the value that is necessary to operate the microcontroller, the transceiver, and sensor circuits of the wireless sensing platform. Consequently, it is necessary to use a dc-dc converter circuit in order to bring the voltage to a desired value and furthermore, regulating circuitry maybe necessary in order to minimize the variation of voltage. All these circuits penalize further the overall power conversion efficiency of the energy harvesting system and must be carefully selected and designed.

Finally, due to the time varying and many times random nature of the available ambient energy, the implementation of energy autonomous circuits for communication and sensing dictates the integration of multiple energy harvesters in order to ensure an average energy supply. In this case, combining the dc outputs of each energy harvesting device must also be done carefully because the efficiency of energy harvesting devices is also dependent on the load that is connected to them and the interconnection of different harvesters that present different and variable loads to each other will also affect efficiency.

These considerations demonstrate on one hand the great challenge for the designer in order to design an energy harvesting assisted wireless sensing platform, but on the other hand, they show the broad nature and the large amount of possibilities that arise by exploring the different disciplines related to the field of energy harvesting.

2 2D-3D Integration for Autonomous Sensors

Sangkil Kim[1]
Pusan National University, South Korea

2.1 Introduction

The advances in energy harvesting technologies for powering low-power sensing platforms are intimately related with low-cost fabrication methods that are compatible with low-cost, flexible substrate materials. Additive manufacturing techniques provide such a platform to fabricate sensors and electronics with low cost, implicitly generate less waste, utilize flexible and low-cost substrates such as paper and plastics, and moreover enable a very quick turnaround, on-demand fabrication and design iteration that facilitates both research and in a way revolutionizes production [16]. In this chapter, we focus on inkjet printing fabrication, an additive manufacturing technique that has shown great potential in the last decade in flexible electronics on both plastics and organic paper substrates, fabricating radio frequency electronic circuits using low-cost [17] and medium-cost equipment [16] even up to millimeter waves. Inkjet printing is suitable for fabricating solar cells, thermoelectric generators, microelectromechanical systems (MEMS) transducers, circuit components such as inductors and capacitors, and transmission lines and antennas with sufficient resolution and provides a platform for packaging and integrating, integrated circuits (ICs), sensors, and interconnects [16]. The recent advances in other additive manufacturing technologies such as 3D printing will undoubtedly help further develop this exciting field of energy harvesting assisted wireless sensor platforms.

The demands for flexible sensors keep increasing as the market rapidly grows for ubiquitous computing, logistics, wearable/implantable electronics, and the Internet of Things (IoT). Flexible sensors have many advantages, such as flexibility, and functionality. They also allow low-cost implementation over some conventional sensors when they are integrated with nanotechnology-based novel materials and cost-efficient fabrication methods such as inkjet printing technology. The flexible sensors can be mounted on rugged/curved surfaces with low cost, which allow them large-scale deployment. The lifetime of the sensor is also an important factor for stable sensing. The energy harvesting and storing technology for flexible sensors is necessary to implement standalone autonomous sensor systems. Selecting proper flexible materials such as paper, liquid crystal polymer (LCP), and polyimide for the substrate material depending on the

[1] Edited by Apostolos Georgiadis.

application is an important step in the implementation of flexible electronics. However, it is more important to choose and understand properly the materials used for sensing functions, electronics, and energy sources as well as fabrication methods that are compatible with a desired flexible sensor application.

Inkjet printing technology is widely utilized and studied as a novel fabrication method compared to the conventional fabrication methods such as milling and wet etching. Numerous electronics utilizing inkjet printing technology such as the Internet of Things (IoT), radio frequency identification tags (RFIDs), and wireless sensor networks (WSNs) have been demonstrated [18, 19, 20]. It is an additive fabrication method that deposits the controlled amount of functionalized ink such as silver nanoparticles, polymers, and nanocarbon structures on a desired position. This technology is cost efficient and environmentally friendly because it doesn't produce any byproducts due to its additive fabrication property. Small feature sizes (less than 50 μm) and arbitrary geometries can be also easily achieved without any masking [21, 22, 23]. Inkjet printing technology has great advantages for implementation of flexible sensors because it is able to print various nanoparticle-based materials, including metals, polymers, and sensing materials. Furthermore, inkjet printing technology can print materials on very thin flexible substrate without damaging the substrate.

Inkjet printing technology has attracted significant interest from many researchers due to the development of numerous types of nanoparticle-based inks such as metals, polymers, and carbon-based materials [24, 25, 26]. The silver nanoparticle inks allow metallization of electronic components and devices using inkjet printing technology, and the development of polymer inks enables printing numerous electronic components like such as transistors, inductors, and capacitors [27, 28]. Also, inkjet printing of nanocarbon materials such as carbon nanotube (CNT) and graphene significantly improved the sensitivity, selectivity, and application spectrum of inkjet-printed flexible sensors [29, 30].

Toward standalone autonomous flexible sensor platforms, energy harvesting is one of the most important design specifications because the available energy affects the system-level design of the sensor platform. There are many types of power sources such as batteries, solar cells, and nanogenerators. The solar cells and batteries can support relatively high power, but they need large area and are not suitable for flexible electronic applications. However, nanowire-based nanogeneratrors [31], for example, are becoming a promising power source for the flexible sensors because they can convert bending motions to power by utilizing, for example, piezoelectric effects. The generated power from the nanogenerator can be stored in printed flexible capacitors, which can be used to operate the sensor platform.

In this chapter, novel materials and fabrication technology for flexible sensors are introduced. The characteristics of nanomaterials including silver nanoparticles, printable polymers, and carbon-based materials such as CNT and graphene are presented, and inkjet printing technology is discussed as a novel fabrication method for flexible sensor system. The state-of-the-art nano-technologies

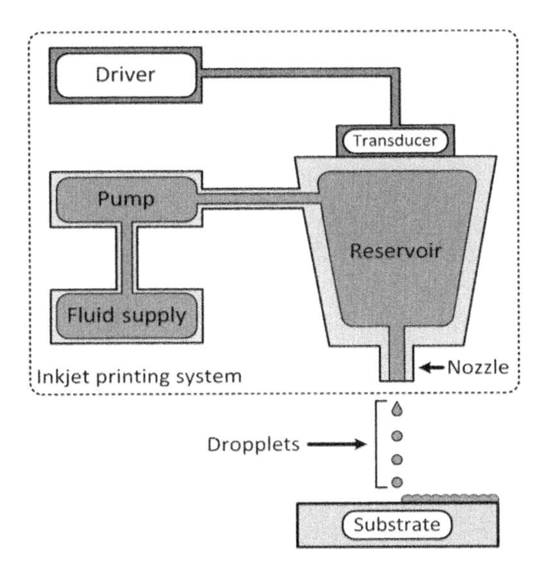

Figure 2.1 Inkjet printing system.

for energy harvesting such as nanowire-based nanogenerators and printed flexible capacitors are also introduced.

2.2 Inkjet Printing Technology

Inkjet printing enables the implementation of printed electronics on various flexible and organic substrates. Inkjet printing is able to print numerous materials such as metals, carbon-based nanostructures, and polymers [24, 25, 26]. It is a cost-efficient, environmentally friendly, and fast fabrication method due to its additive fabrication properties.

2.2.1 Types of Inkjet Printing

The concept of inkjet printing is relatively simple, and it is presented graphically in Figure 2.1. A liquid ink in a reservoir is jetted on a substrate through a nozzle. A desired pattern can be printed by moving the nozzle or substrate to deposit the ink drops on the correct positions.

There are two main inkjet printing methods: the continuous inkjet (CIJ) method and the drop-on-demand (DOD) method, which are shown in Figure 2.2a and 2.2b respectively. CIJ ejects ink drops continuously from the reservoir at a constant frequency (50 kHz–170 kHz), as shown in Figure 2.2c. The ejected drops are charged by charging plates, and the charged drops are directed by a pair of electrodes to print on a substrate or to a gutter for reuse. The advantages of the CIJ method are the high velocity of the ejected ink drops and the high drop ejection frequency. The high velocity of the ink droplets allows

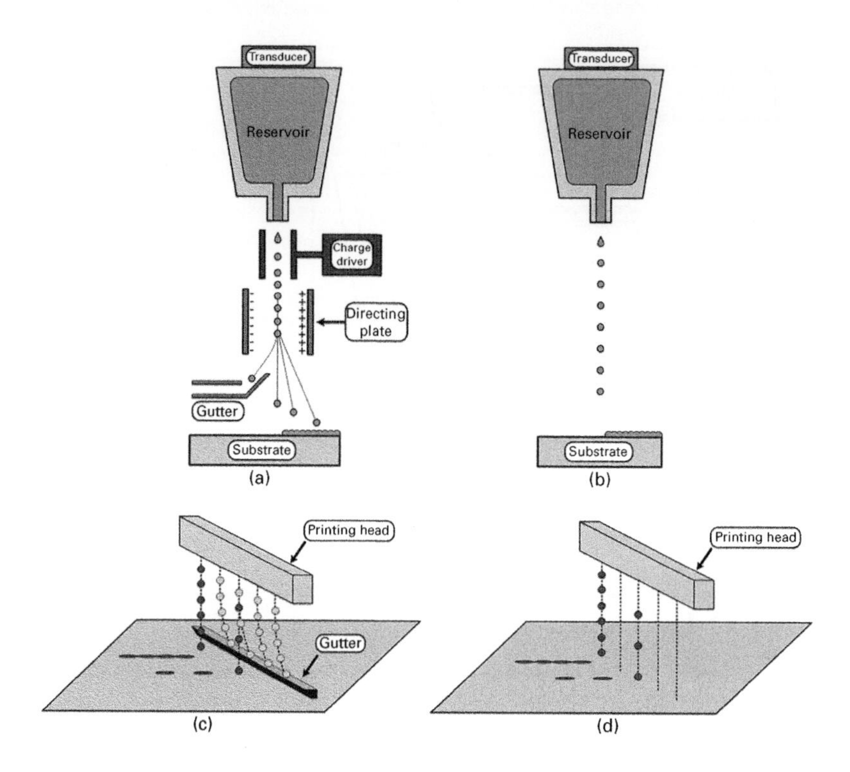

Figure 2.2 Continuous and DOD inkjet printing.

for a relatively long distance between a substrate, and a printing head and the high ejection frequency allows for high-speed printing. The continuous ejection of the ink mitigates the clogging of the printing nozzles. Therefore, volatile solvents such as alcohols can be printed easily. However, the CIJ system requires inks that can be electrostatically charged, and continuous viscosity monitoring is necessary.

DOD printing is similar to CIJ in that a transducer generates a pressure in order to eject a drop of ink. However, the ejected drops are not directed by electrostatic plates. A signal is sent to the transducer, and the transducer ejects the ink drop when it is needed, as shown in Figure 2.2d. DOD printing can use a wider range of inks with varying viscosities and surface tensions, and a higher printing resolution can be achieved compared to the CIJ method. However, clogging of the nozzles happens easily due to the ink drying and the inconsistent use of the nozzles.

There are several types of actuators such as piezoelectric, thermoelectric, and electrodynamic, which generate a pulse in the reservoir in order to eject the ink drop as shown in Figure 2.3. The piezoelectric and the thermoelectric actuators generate the pressure pulse in the reservoir while the electrodynamic actuator creates an ink drop by disturbing the surface tension of the meniscus formed at the end of the nozzle. The thermoelectric actuator generates heat

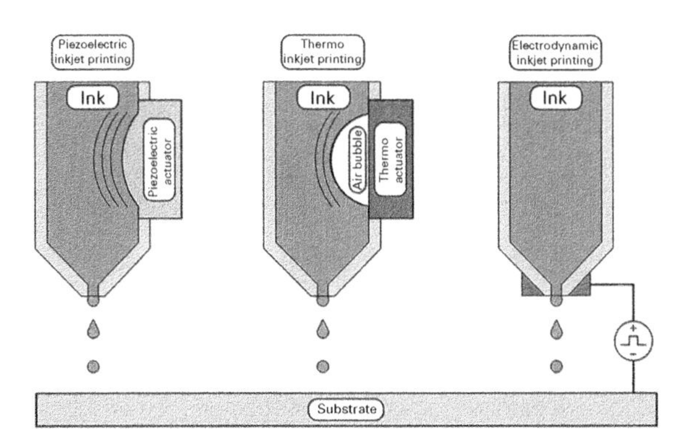

Figure 2.3 Actuator types of the inkjet printing technology.

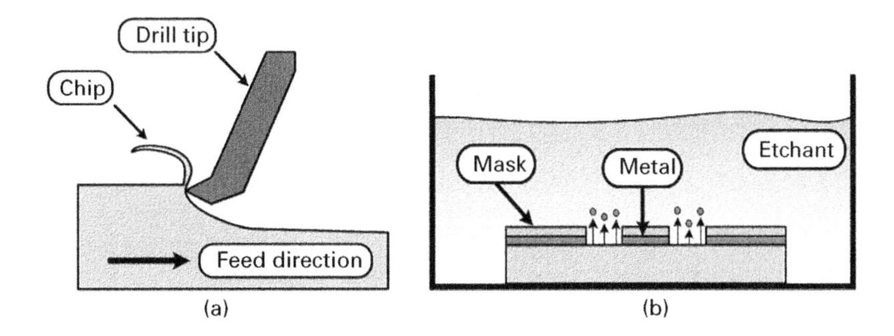

Figure 2.4 Subtractive fabrication methods: (a) milling and (b) wet etching.

causing vaporization of the ink in the reservoir to form a bubble, which ejects a droplet through a nozzle. Small drop sizes and high nozzle density are the advantages of a thermo actuator, but it has limitations on usable ink types. The inks should be able to be vaporized as well as to withstand high temperature. The electrodynamic actuator is able to create an ink drop of a very small size but has disadvantages such as low nozzle density, system complexity, and slow printing speed. The piezoelectric actuator utilizes a piezoelectric material in the reservoir to generate a pressure pulse to eject an ink drop. This type of inkjet printing technology is widely used in the research area because the piezoelectric actuator is compatible with a wider variety of inks.

2.2.2 Inkjet Printing Technology as a Fabrication Method

Inkjet printing technology is an additive method unlike a subtractive method including the wet etching and milling techniques. The wet etching and the milling techniques are widely used fabrication methods due to advantages such as rapid prototyping at low cost. The milling technique cuts and the wet etching technique washes away the unwanted materials selectively, utilizing a milling machine or an etchant as shown in Figure 2.4.

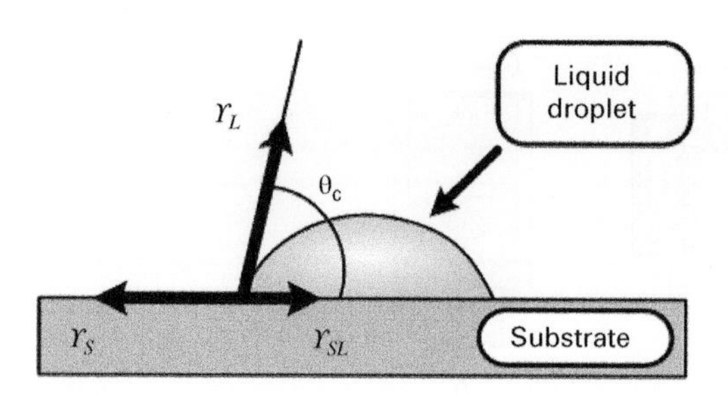

Figure 2.5 Surface energy.

The milling technique requires a milling machine, substrate, and cutting bits that have a size equal to or smaller than the smallest feature size of the desired pattern. A variety of features such as holes and slots can be created by the milling process, and this process is compatible with lots of materials, including metals and ceramics with tolerance down to 25 μm. However, a lot of byproducts are formed like chips and rough edges because of the cutting of the drill bit (Figure 2.4a). This method is suitable to process hard and relatively thick materials, but can be hardly used on thin flexible substrates as the bit removes part of the substrate. The wet etching technique is also commonly used to fabricate a printed circuit boards (PCBs). The etchants are solvent containing highly corrosive acids that dissolve a metal or a substrate (Figure 2.4b). The used etchants combined with the discarded materials form waste produced by the fabrication process, and they require special treatment for safety and environmental protection reasons.

Inkjet printing technology is a more efficient and environmentally friendly fabrication method compared to those conventional fabrication methods. It drops ink on an exact desired position. Therefore, there are no wasted materials because the inkjet printing produces no byproducts such as the acid etchant. In addition, a reasonably high resolution down to 50 μm can be achieved with high repeatability and without any special surface treatment such as masking. It is possible to improve the printing resolution to submicrometer values [23]. Furthermore, inkjet printing is compatible with very thin or flexible substrate because it doesn't damage the substrate surface.

2.2.3 Inkjet Printing and Surface Energy

Each substrate has different physical surface properties such as roughness and surface energy that result in different inkjet printability [22, 32]. The surface energy of the substrate is a very important factor because it determines substrate wetting. The difference in surface free energy between the printed ink drop and the substrate determines whether the ink wets the substrate surface (spread on the substrate) or not (forms a ball on the surface). It can be defined by the contact angle θ_c as shown in Figure 2.5.

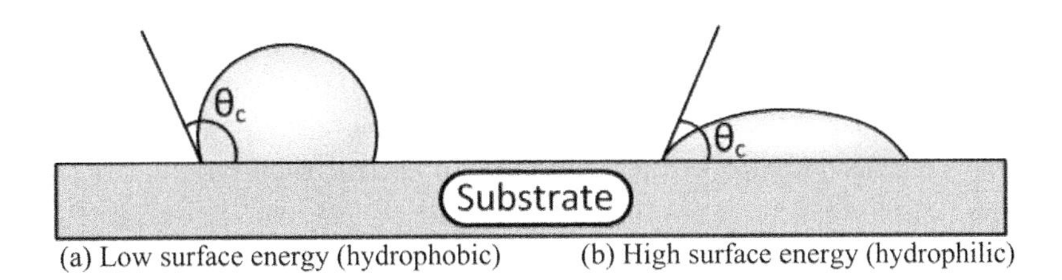

Figure 2.6 Low and high surface energy.

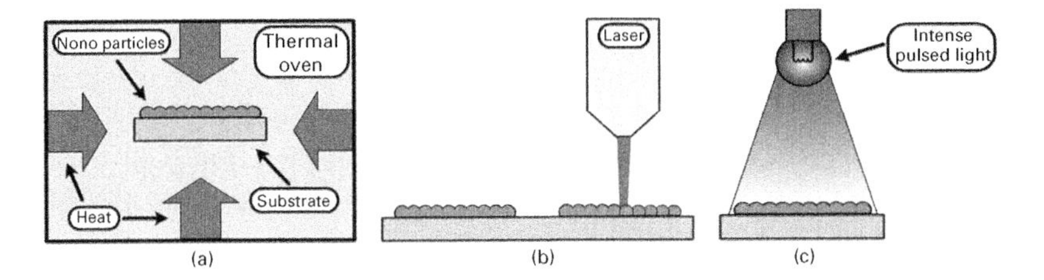

Figure 2.7 Sintering processes: (a) thermal sintering, (b) laser sintering, and (c) UV plasma sintering.

Young's equation (2.1) describes the balance of forces caused by a liquid droplet on a dry surface that results in a one-dimensional force equilibrium along the solid boundary

$$\gamma_S = \gamma_L \cdot \cos\theta_c + \gamma_{SL}, \tag{2.1}$$

where γ_S is the solid surface energy, γ_L is the liquid surface energy, and γ_{SL} is the solid/liquid interfacial surface energy. A hydrophobic surface has a high contact angle $\theta_c > 70°$C that indicates a low surface energy γ_S, while a hydrophilic surface has a low contact angle $\theta_c < 30°$C, which indicates a high surface energy γ_S as shown in Figure 2.6. Too high of a contact angle θ_c is hardly able to form a continuous printed layer, and too low of a contact angle θ_c results in drop spreading.

2.2.4 Sintering Process

A sintering process is necessary to evaporate the solvent and make the printed silver nanoparticles conductive. The conductivity of the inkjet-printed metal is dominated by the sintering process. Therefore, it is very important to understand the properties of sintering processes and select the proper sintering process for each application. Many kinds of sintering processes are used, such as thermal, laser, and ultraviolet (UV) plasma sintering [33, 34, 35, 36].

The most widely used sintering process is thermal sintering. It is relatively simple and easy to apply to inkjet-printed nanoparticles. This process heats both the printed nanoparticles and substrate together as shown in Figure 2.7a.

A sintering duration, a temperature ramping ratio, and a sintering temperature are the most important parameters of thermal sintering. Usually the inkjet-printed silver nanoparticles require about one hour of sintering at a temperature of 180°C to get a saturated conductivity value. A longer sintering interval is required when a lower sintering temperature is utilized.

The ramping ratio is another important factor necessary in order to make a continuous and uniform printed layer from the inkjet-printed nanoparticle inks. For example, the inkjet-printed silver nanoparticles can't form a continuous even surface if the ambient temperature changes too rapidly. If the temperature ramps too fast, cracks on the metal layer are produced. The sintering temperature is the most important factor that determines the quality of the inkjet-printed nanoparticles and sintering duration of the inkjet-printed nanoparticles. For instance, the conductivity of the inkjet-printed silver nanoparticles is a function of the sintering temperature, and less time is required as the sintering temperature increases. However, it should be noted that sintering temperature and time should be selected after taking account of the substrate's temperature durability as well as the desired conductivity of the printed silver nanoparticles. The thermally sintered silver nanoparticles have a resistivity of about 180 $\mu\Omega$·cm–8.3 $\mu\Omega$·cm depending on sintering temperature and printed number of layers, which is about 5.2–112.5 times of that of a bulk silver (1.6 $\mu\Omega$·cm) [36].

Laser sintering utilizes a laser to sinter the printed nanoparticles at room temperature. Thermal sintering may damage the substrate due to the high heat required to melt nanoparticles together while the laser sintering is able to sinter the inkjet-printed nanoparticles selectively as shown in Figure 2.7b. This process heats the printed nanoparticles at very high temperature with very little heating of the substrate, and consequently it requires shorter time than the thermal sintering process. The laser sintering is sometimes applied to the thermally sintered silver nanoparticles to increase the conductivity. A laser power, a scanning speed, and a radius of the laser focus are the most important parameters in this process. The laser power and scanning speed should be adjusted so that the printed nanoparticles absorb the majority of the heat from the laser. The radius of the laser focus determines the resolution of the sintering process. The reported laser-sintered silver nanoparticles have a resistivity of about 9.1 $\mu\Omega$·cm which is about 5.7 times that of a bulk silver [36].

UV plasma sintering is the fastest sintering processes among the various sintering processes at room temperature without damaging substrate, and this method is briefly described in Figure 2.7c. The printed nanoparticles are exposed to an intense pulsed flash light that has a broad spectrum in visible range. This process takes only a few milliseconds, and the intense pulsed light is generated by an arc plasma phenomenon in the flash lamp. The UV sintered silver nano-particles have a resistivity of about 4 $\mu\Omega$·cm–8 $\mu\Omega$·cm, which is about 2.5–5 times that of bulk silver [34].

2.3 Nanomaterials

2.3.1 Silver Nanoparticles

Among the numerous metal nanoparticle inks, the silver nanoparticle ink is one of the most widely used ink for the inkjet-printed conductor because of its relatively high conductivity and low sintering temperature compared to copper (Cu) and gold (Au) ink. The majority of printed nanoparticle inks are not conductive before the sintering process because the printed silver nanoparticles are coated with a polymer, which helps maintain the nanoparticles in ink form. The sintering process is necessary to make the printed nanoparticles conductive since this process burns off the polymers and impurities in the solvent. Moreover, the bonding strength of the printed silver traces with the substrate is increased, and the nanoparticles create a percolation channel for the electrons. The conductivity of the printed silver nanoparticles is affected by the numbers of printed layers, sintering temperature, and nanoparticle concentration of the ink.

The surface of deposited inkjet-printed silver nanoparticles is shown in Figure 2.8a, where a Veeco Atomic Force Microscope (AFM) has been utilized to scan the surface of an inkjet-printed sample [37]. The measured arithmetic average R_a is about 11.4 nm while the root mean squared R_q roughness is about 14.4 nm, as shown in Figure 2.8b. The mechanical and electrical properties of the inkjet printed nanosilver particles are thoroughly studied in [32, 38]. The pull-off breaking force of the printed silver nanoparticles is about 50 N, while Young's modulus of the thermally sintered silver trace is a function of the sintering temperature.

The water contents of the substrate affect on the adhesion strength of the printed traces to substrate rather than conductivity of the printed conductors [32]. The sheet resistances of the printed traces with different printed layers at different sintering temperature are measured using Cascade's four-point probe station as shown in Figure 2.9a. The sheet resistance decreases when the number of printed layers and the sintering temperature increase. It is because the high sintering temperature helps to form a good channel for electron flow and adding more layers increases the particle density, which results in an uniform solid structure. The conductivity σ can be extracted using the cross-section area A, the length l, and the resistance R of the printed patterns of the trace as shown in (2.2):

$$\sigma = \frac{l}{A} \cdot \frac{1}{R}. \tag{2.2}$$

The extracted conductivities are shown in Figure 2.9b. The maximum conductivity value of the inkjet-printed silver nanoparticles is $1.2107 \cdot 10^7$ S/m, which is about 18.75% of silver's bulk conductivity ($6.4 \cdot 10^7$ S/m). This value is almost the same as the conductivity of bulk iron ($1.04 \cdot 10^7$ S/m). It suggests that the inkjet-printed nanoparticles can be used for implementing microwave devices such as RFIDs and sensors.

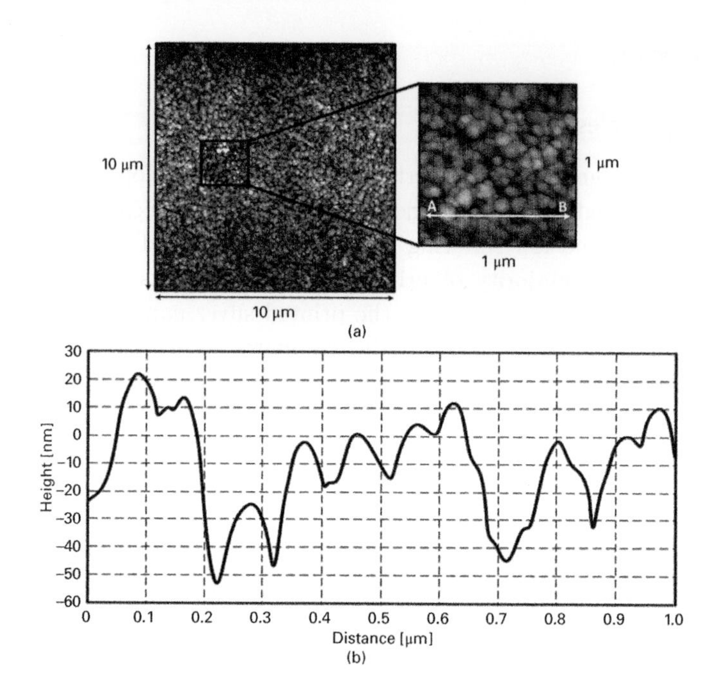

Figure 2.8 (a) The surface of the inkjet-printed nanosilver ink. (b) Cross section of the line AB. ©2013 IEEE. Reprinted, with permission from [37]

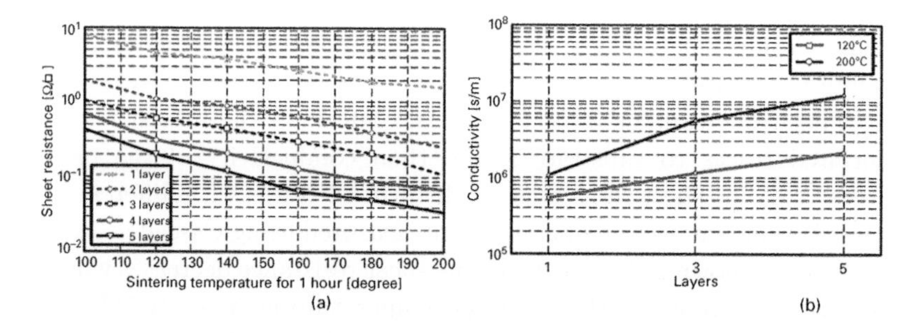

Figure 2.9 (a) Sheet resistance and sintering temperature and (b) extracted conductivity of the printed silver traces [36].

2.3.2 Inkjet-Printable Polymers

Polymers are usually flexible materials and form compounds of repeating structural units such as a plastic and a polyimide. Polymers have numerous applications for flexible electronics such as a spacer, an insulator, or a sensing material. The importance of polymers in electrical engineering has critically increased as the demands for flexible electronics increase. Recently, inkjet-printable polymer inks have been developed, and their performance is reported [39, 40, 41, 42]. Polymer-based dielectric inks utilizing SU-8 ($\epsilon_r = 4$) [43], polyvinylpyrrolidone (PVP) ($\epsilon_r = 3$) [44], and polymethyl methacrylate (PMMA) [45] inks are

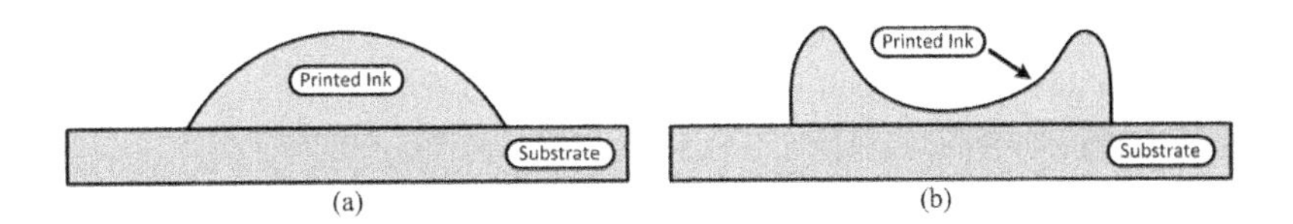

Figure 2.10 Inkjet-printed ink: (a) normal ink droplet and (b) coffee ring effect.

formulated for printing multilayer electronic components such as capacitors. Those polymers are widely used because they have a strong chemical resistance after the printing and sintering processes. The inkjet printing process of the polymer is similar to that of silver nanoparticles. Similarly with the sintering process, a cross-linking process is necessary after printing polymer inks in order to link polymer chains.

There are two types of cross-linking methods: heat and UV cross-linking. The heat cross-linking process consists of heating the printed polymers in the oven at a high temperature (usually higher than 180°C). The UV cross-linking process consists of exposing the printed polymers to a UV flash light.

SU-8 is a well-known photoresist and it is suitable for inkjet printing. The low-temperature UV cross-linking process is compatible with the inkjet-printed SU-8, and a low viscosity can be maintained with high polymer content by weight [46]. PVP is a polymer that is used as an insulation layer for field effect transistors (FETs), and inkjet-printed PVP has been reported in [40]. PVP ink has a higher viscosity at a low polymer concentration by weight in a solvent. PMMA is another commonly used polymer as a display device and a dielectric layer for transistors [47], and it has been demonstrated to be inkjet printable [45].

A coffee ring effect is a common problem of inkjet printing technology, especially in the inkjet printing of polymers as shown in Figure 2.10 [38]. This effect results from higher evaporation flux at edges of inkjet-printed patterns, and capillary force drives flow of liquid to the edges to compensate for the evaporation losses. The coffee ring effect can be suppressed by controlling evaporation of the solvent by combination of a low and high boiling solvent [48, 49]. In this way, the coffee ring effect can be suppressed and a homogeneous polymer film can be formed.

2.3.3 Nanocarbon-Based Materials (Graphene and Carbon Nanotubes – CNTs)

Carbon nanomaterials such as CNT and graphene are a very promising research area for electronic devices. Both CNT and graphene consist of latices of carbon atoms, but their geometries are different. CNT has a cylindrical geometry while graphene has a planar geometry. The diameter of a CNT is in nanometer scale, while the length of the tube is in the micrometer scale, which results in a very high length-to-diameter ratio. Both nanocarbon materials have unique electrical and mechanical properties. Especially graphene has lots of promising properties such as high electron mobility and thermal conductivity of 2,105 $cm^2V^{-1}s^{-1}$ and 5,103 $Wm^{-1}K^{-1}$, respectively. Both CNTs and graphene are used to implement gas sensors or electronic components such as transistors due to their high reactivity to gas molecules and electron mobility. Gas sensors utilize the impedance change of the CNTs and graphene at high frequency when they are exposed to gases such as a carbon dioxide (CO_2) or ammonia (NH_3) [50, 51, 52]. The selectivity of the nanocarbon-based sensors can be improved when the nanocarbon structures are functionalized to the specific gas molecules [53]. Also, electronic components such as diodes and transistors are reported utilizing CNTs and graphene [54, 55, 56], where nanocarbon materials are utilized as a channel of a Schottky diode [55] or a transistor [54, 56].

Nanogenerator technology is aiming to generate the green and sustainable energy for small electronic systems such as mobile and portable electronics. Many types of nanogenerators are reported and demonstrated using piezoelectric effects. Nanotechnology-based capacitors are also actively studied because capacitors are one of the most important electronic components in energy harvesting technology. The advances in inkjet-printing technology have led to significantly improved performance of nanotechnology-based inkjet-printed capacitors. Inkjet-printed capacitors don't require the sequence of photolithography and etching steps, although the capacitors are multilayer structure (metal–insulator–insulator, MIM) because inkjet-printing technology is a purely additive fabrication process. In the next section, nanowire-based nanogenerators utilizing piezoelectric and pyroelectric effect are introduced and nanotechnology-based capacitors for energy harvesting are presented.

2.4 Nanowire-Based Piezoelectric Nanogenerators

The piezoelectric effect is the generation of electron charges in certain solid materials such as crystals and ceramics when they experience mechanical stresses such as pressure or vibration, as shown in Figure 2.11. This effect was discovered by French physicists Paul-Jacques Curie and Pierre Curie in the 1880s, and it is widely used in numerous applications nowadays such as a sound detector, an electronic clock generator, and a voltage generator [57]. In 2006, the first zinc oxide (ZnO)-based nanowire nanogenerator has been reported using the piezoelectric effect [58]. The ZnO nanowire generated a voltage or a current when an AFM tip sweeps across the ZnO because of the coupling between the piezoelectric and semiconducting properties of the ZnO nanowire, as shown in Figure 2.12a. The proposed design generates an open-circuit voltage of 9 mV. In 2008, ZnO nanowire was bonded horizontally on a flexible substrate and an AC electric energy was generated by bending the substrate as shown in Figure 2.12b [59]. A single ZnO wire with a diameter of 4 μm and a length of 200 μm on flexible kapton substrate generated 20 mV–50 mV and 400 pA–750 pA respectively. As another example, an integrated ZnO nanowire-based nanogenerator has been

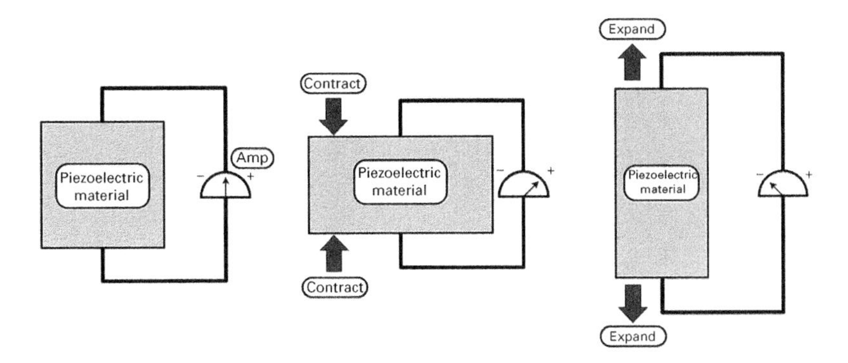

Figure 2.11 Piezoelectric effect.

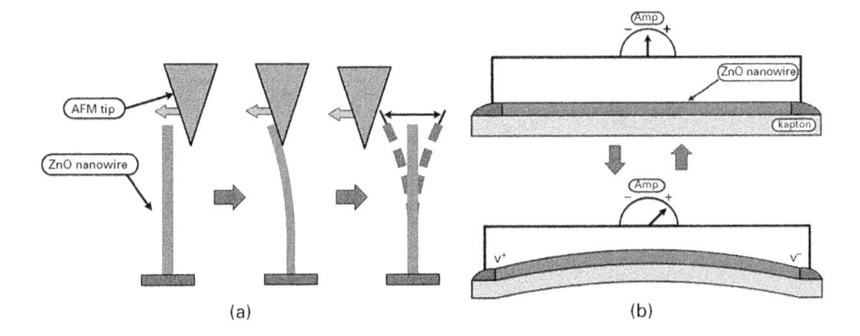

Figure 2.12 (a) The resonance vibration of a nanowire after being released by the AFM tip [58] and (b) bending of the substrate and a piezoelectric potential [59].

reported that obtained a peak open-circuit voltage of 37 V and a peak short-circuit current of 12 μA using an 1 cm^2 of nanowire array [60].

2.5 Nanotechnology-Based Capacitors

A capacitor is an electronic component that saves electrical energy. There are numerous types of capacitors such as ceramic, electrolytic, and mica capacitors on the market. A capacitance C of a capacitor is the ability of the capacitor to store an electric charge. Basically, a capacitor can be briefly modeled as a parallel plate capacitor, as shown in Figure 2.13. Its capacitance can be written as

$$C = \epsilon_0 \epsilon_r \frac{A}{D}, \tag{2.3}$$

where ϵ_0 is the vacuum permittivity ($\epsilon_0 = 8.854 \cdot 10^{-12}$ F/m), ϵ_r is the dielectric constant (or relative permittivity), A is the area of the metal plate, and D is the

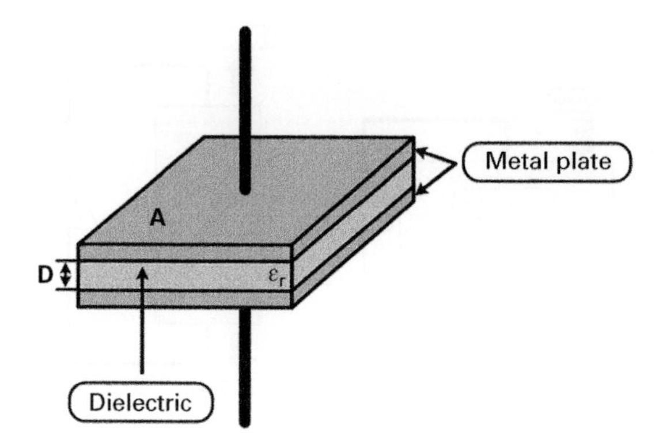

Figure 2.13 Parallel plate capacitor.

distance between the metal plates. The electrical energy (E) stored in a capacitor of capacitance C is equal to

$$E = \frac{1}{2}CV^2, \tag{2.4}$$

where V is a voltage across the capacitor and the capacitor can store electrical energy until it experiences breakdown that is determined by the dielectric material. The stored electrical energy can be used as a source like a battery until it releases all of the stored electrical energy.

Low-leakage capacitors have some important advantages over batteries, and consequently they are used to collect and store harvested ambient RF energy like, for example, that from TV and radio signals [15]. The number of charge–discharge cycles of capacitors is higher than that of batteries, and they present low internal thermal losses that are expressed by an equivalent series resistance (ESR). Therefore, capacitors can source or sink larger amounts of charge compared to a battery. Furthermore, capacitors are very stable because they are not volatile when exposed to harsh environments, and there is no degradation from shallow discharge like, for example, with nickel-cadmium (NiCAD) batteries. Plus, capacitors are more environmentally friendly because they don't utilize heavy metals and they do require a dedicated waste disposal process like batteries do.

The advances in nano- and inkjet printing technology enabled the implementation of inkjet-printed flexible capacitors [39, 40, 41, 42]. Polymer-based inks such as SU-8 photoresist and PVP are developed and inkjet-printed as an insulator layer between two metal plates. Those two polymers are widely used because they have strong chemical resistance after cross-linking.

Flexible inkjet-printed capacitors that have a self-resonant frequency (SRF) around 3 GHz have been demonstrated in [42]. Dielectric inks made from SU-8 and PVP were printed between silver nanoparticle ink electrodes on a flexible polyimide substrate. A photo of the PVP-based prototypes is shown in Figure 2.14. The area of the printed capacitors was 1.5 × 1.5 mm^2 and the

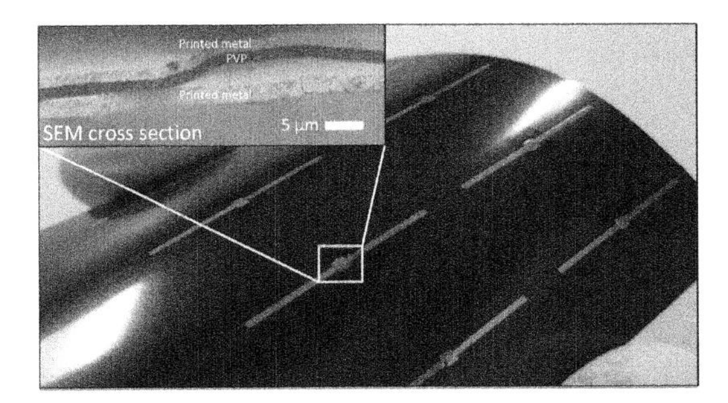

Figure 2.14 Inkjet-printed parallel plate capacitors on polyimide. ©2013 IEEE. Reprinted, with permission from [42]

thickness of the inkjet-printed SU-8 was 4 μm while that of the inkjet-printed PVP was 0.8 μm. The capacitance of the SU-8 based capacitor was about 20 pF and it had a self-resonance frequency around 3 GHz, while the capacitance of the PVP-based capacitor was about 50 pF for the same metal area with a self-resonance frequency of 1.9 GHz. The SU-8-based capacitor has a higher self-resonance frequency because of the larger thickness of the inkjet-printed polymer dielectric than that of the PVP-based capacitor. The maximum Q factor value was approximately 4 due to the thin metal layers ($\sim$ 1.5 μm) and the step-discontinuity at the edges of the dielectric.

2.6 Problems and Questions

1. What are the drop-on-demand (DOD) and continuous inkjet printing (CIJ) methods?

2. Why is a sintering process required after printing nanoparticle ink?

3. What are the thermal, laser, and UV plasma sintering processes?

4. What are the piezoelectric and pyroelectric effects?

5. Find the capacitance of the parallel plate capacitor shown in Figure 2.15 (do not consider any fringing effect and let $\epsilon_o = 8.8510^{-12}$ F/m). Find the stored energy when a voltage of 5 V is applied to this capacitor.

6. Find the conductivity (σ) of a printed line shown in Figure 2.16. The profile follows a curve defined by the equation $y = -0.0048x(x-1)$, the length (l) of the line is 10 mm, and the resistance (R) of the trace is 2.0 Ω.
7. Which one of the two surfaces (a) and (b) shown in Figure 2.17 is hydrophilic and which one is hydrophobic?
8. What is the coffee ring effect and how can we suppress this phenomenon?

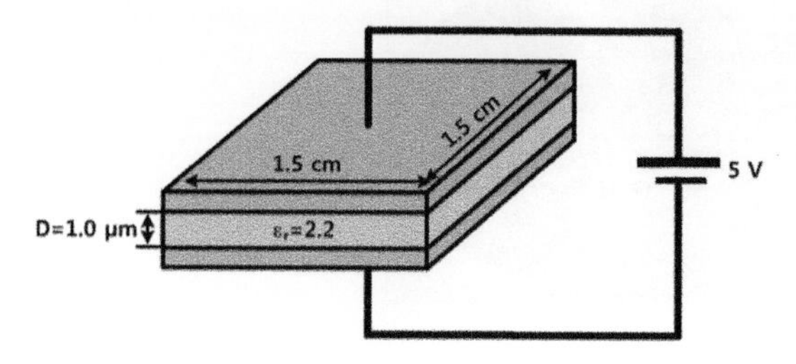

Figure 2.15 Parallel plate capacitor.

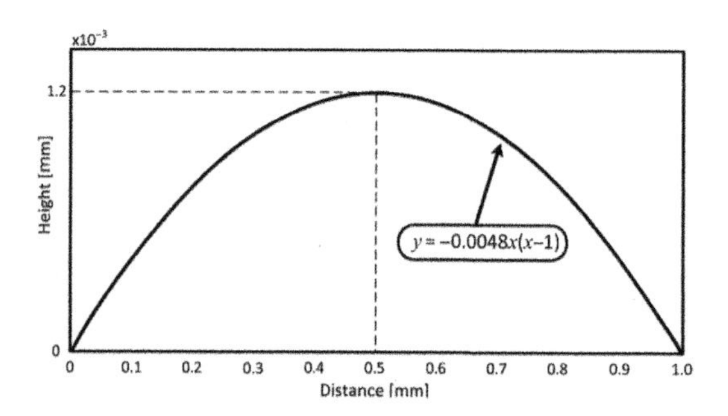

Figure 2.16 Cross-sectional profile of a printed line.

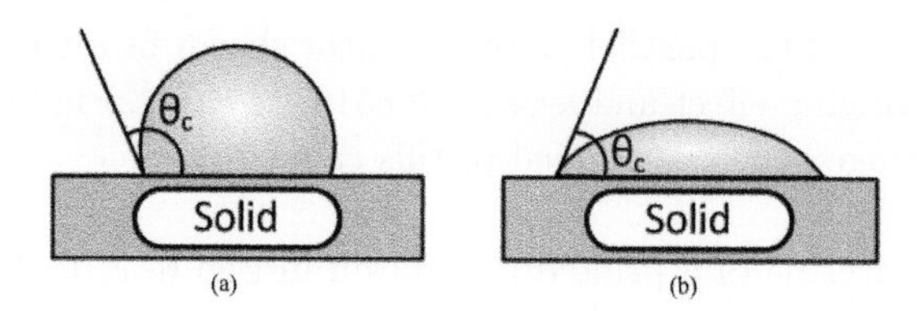

Figure 2.17 Hydrophylic and hydrophobic surfaces.

3 Solar (Light) Energy Harvesting

3.1 Introduction

The ubiquitous presence of sunlight makes solar energy one of the most abundant sources of ambient energy. Consequently, harvesting of light wave energy is one of the most important applications of energy harvesting technologies. This chapter presents the principles of solar cell operation and focuses on the challenges and optimization of light energy harvesting circuits and their integration with other harvester circuits and antennas, topics that are fundamental toward the implementation of energy autonomous wireless sensors.

The chapter begins with an introductory section describing the origins of solar cell technology. Next measures of light energy and the solar cell model and efficiency are presented. Finally, the integration of solar cells with antennas and the combination of various energy harvesting sources are discussed.

3.2 History

Solar energy harvesting, or more generally light energy harvesting, consists of converting energy contained in photons of light into electrical energy. This is a one-step process where a photon, a quantum of energy of light, is absorbed by an electron in a solid. The electron is excited in a higher energy state where it is able to move, generating an electric current [61].

This phenomenon was first studied by Edmond Bequerel in 1839 [62]. Bequerel observed that an electrical current was produced when two platine or gold electrodes dipped in a solution that can be acid, neutral, or alcaline are illuminated by unequal light intensity. This is known as the photovoltaic effect. Experiments by other researchers followed and, most notably, in 1876, William Adams and Richard Day observed an electrical current when a selenium sample in contact with two heated platinum contacts was illuminated by light [61]. The first solar cell is attributed to Charles Fritts, who installed the first rooftop solar panel by coating selenium with gold in 1894.

When a semiconducting material such as silicon is illuminated by light, light photons with energy above the band gap of the material excite carriers, which results in an observed electrical current [6]. Specifically, electrons from the valence band of the semiconductor that absorb the incoming photons are excited into the conduction band, resulting in electron–hole pairs and therefore an increased conductivity. A piece of semiconducting material with ohmic contacts is a photoconductor. In order to be able to generate current and avoid the subsequent fast recombination of the electron–hole pairs, photodiodes are utilized instead of photoconductors [63]. There exist four types of photodiodes, namely p-i-n, p-n, heterojunction, and metal-semiconductor (Schottky barrier) photodiodes [64]. Conventional wafer type solar cells are p-n photodiodes, whereas thin-film solar cells are typically p-i-n photodiodes [63].

There are three generations of solar cells. The first generation includes wafer-based solar cells mainly built on silicon. The disadvantages of wafer-based solar cells are both high cost and low solar-to-electrical energy conversion efficiency. Approximately half of the cost of a Si wafer (c-Si) photovoltaic module is attributed to wafer preparation [63]. The second generation began in the early 1980s and includes thin film solar cells [65]. Thin film technology both reduced material costs and allowed a much larger size of the unit of manufacturing of approximately 100 times compared to the first-generation unit size, which was limited to the wafer size. In 2003, the production cost of first-generation technology was 150 USDm^{-2} with obtained efficiencies of 20%, whereas second-generation technology offered reduced production costs of 30 USDm^{-2} with lower efficiencies of 5%–10% [65].

Third-generation photovoltaic technology combines the low-cost fabrication of thin film technology with novel design concepts able to lead to much higher efficiencies. One such possibility is the use of tandem solar cells where two or more cells of different materials and band-gap energy are stacked on top of each other aiming to maximize the efficiency of photovoltaic conversion of photons with different energy [65]. Another approach is the possibility of exciting multiple electron–hole pairs from high-energy photons. Finally, another technique is the solar thermal electric one, where sunlight is first converted to heat and subsequently converted to electrical energy or reradiated as light (thermophotovoltaics).

Finally, an emerging solar cell technology is based on quantum dots. Quantum dots are semiconductor nanoparticles that have optoelectronic properties that are tunable according to their shape and size. Colloidal quantum dots are synthesized into thin films from liquid solutions, and they have shown a very promising potential in solar cell applications due to the ability to tune their band gap by modifying the fabrication conditions [66] and due to the possibility of large-scale fabrication using, for example, spin coating or inkjet printing methods [67]. They have been considered for single junction or tandem solar cell configurations exploring both the infrared and visible solar spectra [66].

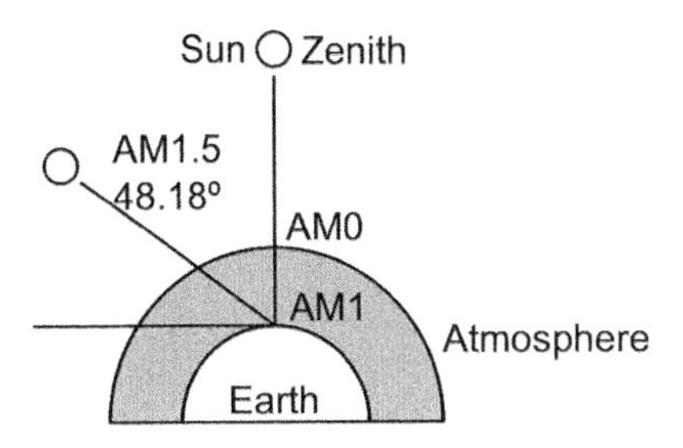

Figure 3.1 Air mass standards.

3.3 Light Sources and Measures

The various light sources are characterized by measuring their irradiance I_λ, which has units of power density (Wm^{-2}). Solar light with a power density of 1,000 Wm^{-2} (= 100 mWcm^{-2}) is also called 1 sun. The irradiance measures the power density of a radiated wave across the complete frequency spectrum. In contrast, the spectral irradiance consists of the irradiance per wavelength, and it is typically measured in $\mathrm{Wm}^{-2}\mathrm{nm}^{-1}$.

The irradiance should be distinguished from illuminance, which is a photometric measure with units lux or lumens per square meter (lux = lm m^{-2}), which measures the perceived power density of light weighted by the sensitivity of the human eye [68]. The sensitivity of the human eye is defined by a luminosity function. The visible light spectrum extends from 380 nm to 750 nm, while the solar spectral irradiance is typically provided in the 280 to 4,000 nm interval.

The American Society for Testing and Materials (ASTM) has developed the standard solar spectral irradiance distributions that are used in evaluating solar cell performance [69]. The air mass zero (AM0) or extraterrestrial (ETR) irradiance spectrum is provided in the standard ASTM E-490-00 developed in 2000 [70] based on several sets of measured data from satellites or space missions. The total solar irradiance of AM0 is 1.366 KWm^{-2}, known as the solar constant [69]. Additionally, ASTM defines two terrestrial solar irradiance distributions, the AM1.5 Global (AM1.5G) and the AM1.5 Direct and Circumsolar (AM1.5D), included in the standard ASTM-G-173-03 [70]. Both distributions are defined assuming a surface that is receiving the solar radiation that has a tilt of 37° toward the Earth's equator. The tilt angle selected is approximately the average latitude for the contiguous U.S.A. The air mass AM quantity is derived from the length of the path that the solar radiation waves are passing through within the earths atmosphere. An air mass of 1 (AM1) corresponds to a solar zenith angle of 0°. The AM1.5 standard distributions correspond to a solar zenith angle of 48.18° as shown in Figure 3.1.

The global spectral irradiance distribution includes both direct and diffuse solar radiation and has a total irradiance of 1 sun = 1 KWm^{-2}. The direct and circumsolar distribution does not include diffuse radiation, and it corresponds to

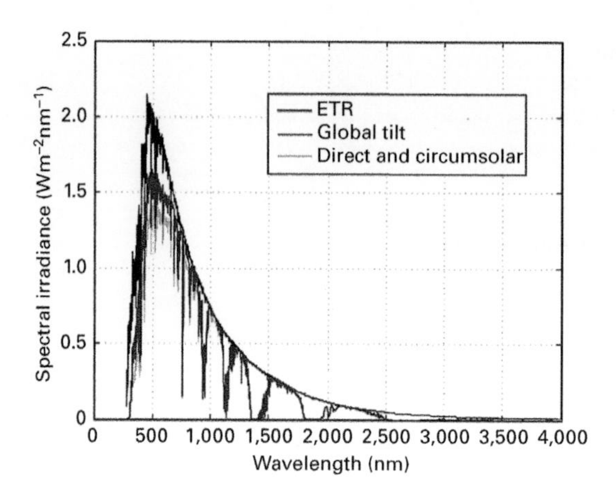

Figure 3.2 Standard solar spectral irradiance distributions.

an irradiance of 0.9 KWm^{-2} [7]. The three standard distributions are shown in Figure 3.2.

Indoor light sources have a much lower irradiance between 100 and 1,000 times less than outdoor sunlight. Specifications for indoor light sources are usually given in photometric units, with a typical minimum value of 500 lux required for libraries and offices [8]. The AM1.5 global spectrum corresponds to 100 Klux in photometric units. However, the actual levels of solar irradiance can vary significantly from the values specified in the AM1.5 standards due to many factors such as the daily variation of the air mass that solar radiation has to travel through during a single day as the position of the sun varies relative to the earth, and due to the presence of clouds that scatter direct sunlight. It is reported in [71] that the average solar irradiance on earth is 170 Wm^{-2} which is approximately one sixth of the AM1.5G standard irradiance.

3.4 Efficiency of Solar Cells

The most important parameter in characterizing the performance of solar cells is their efficiency. The efficiency η is defined as the ratio of the electrical power generated by the solar cell P_o to the incident light power P_i.

$$\eta = P_o/P_i \tag{3.1}$$

The incident power is the integral of the irradiance I_λ over the surface A of the solar cell, which for a uniform irradiance is given by

$$P_i = I_\lambda A. \tag{3.2}$$

The theoretical calculation of the output power of the solar cell requires the application of solid-state theory and thermodynamics. There are several theoretical works dealing with the efficiency of solar cells; however, a widely used model was proposed by Shockley and Queisser [72].

In computing the efficiency of solar cells, Shockley and Queisser assumed that the source of light energy, the sun, and the solar cell itself behave as black bodies with temperature T_s and T_c respectively. In [72], it was postulated that the solar cell efficiency depends on four parameters, namely the energy band gap E_g of the material comprising the solar cell, the ratio between the temperature of the sun T_s and the temperature of the solar cell T_c, the probability t that a photon with energy above the band gap will excite an electron to the conduction band and generate an electron–hole pair and a geometric factor f_w related to the angle subtended by the sun toward the solar cell. This is expressed as

$$\eta = \eta(x_g, x_c, t, f_w), \tag{3.3}$$

where

$$x_g = E_g/kT_s \tag{3.4}$$

$$x_c = T_c/T_s \tag{3.5}$$

and $k = 1.38064852 \cdot 10^{-23}$ m^2 kg s^{-2} K^{-1} is Boltzmann's constant. In (3.3) the normalized energy parameters x_g and x_c were used, similarly to [72]. Shockley and Queisser computed the ultimate efficiency limit and the detailed balanced efficiency limit for p-n junction solar cells, which will be described next. In the following, the probability t is assumed to be equal to one.

3.5 Ultimate Solar Cell Efficiency

The ultimate solar cell efficiency η_g depends only on the energy band gap E_g of the material comprising the solar cell and the temperature of the sun T_s expressed through the single parameter x_g [72].

Every photon with energy $E = h \cdot f > h \cdot f_g = E_g$, where f is the frequency of the photon and $h = 6.62607004 \cdot 10^{-34}$ m^{2}kg s^{-1} is Planck's constant, excites a single electron to the conduction band. Because the sun is considered a black body with temperature T_s it radiates photons whose energy follows the Planck distribution.

The number of photons Q with energy between two limits E_1 and E_2 for a black body described by a Plank distribution with temperature T, is calculated by integrating the Planck distribution from E_1 to E_2,

$$Q(T, E_1, E_2) = \frac{2\pi}{h^3c^2} \int_{E_1}^{E_2} \frac{E^2 dE}{e^{E/kT} - 1} \tag{3.6}$$

Therefore, the number of photons Q_s with energy larger than E_g takes the form $Q_s = Q(T_s, E_g, +\infty)$, which is computed as

$$Q_s = Q(T_s, E_g, +\infty) = \frac{2\pi}{c^2} \int_{f_g}^{\infty} \frac{f^2 df}{e^{hf/kT_s} - 1} = \frac{2\pi (kT_s)^3}{h^3 c^2} \int_{x_g}^{\infty} \frac{x^2 dx}{e^x - 1}. \tag{3.7}$$

It is further assumed that the excited electrons quickly lose any excess energy above E_g as thermal energy. The total output power available by the solar cell is then

$$P_o = hf_g Q_s A. \tag{3.8}$$

The irradiance I_λ of the light impinging the solar cell is given by integrating the Planck distribution from 0 to $+\infty$,

$$I_\lambda = \frac{2\pi h}{c^2} \int_0^{\infty} \frac{f^3 df}{e^{hf/kT_s} - 1} = \frac{2\pi (kT_s)^4}{h^3 c^2} \int_0^{\infty} \frac{x^2 dx}{e^{hx} - 1} = \frac{2\pi^5 (kT_s)^4}{15 h^3 c^2} = \sigma T_s^4, \tag{3.9}$$

where $\sigma = 5.670367 \mathrm{J\ m^{-2} s^{-1} K^{-4}}$ is the Stefan–Boltzmann constant. Using (3.2), (3.8), (3.7), and (3.9), one gets

$$\eta_g = \frac{E_g Q_s}{I_\lambda} = \frac{hf_g Q_s}{I_\lambda} = \frac{15 x_g}{\pi^4} \int_{x_g}^{\infty} \frac{x^2 dx}{e^x - 1}. \tag{3.10}$$

One can easily compute numerically (3.10) for different energy band-gap values and a given sun temperature. Shockley and Queisser showed that the ultimate efficiency η_g has a maximun value of 44% corrseponding to a normalized band gap of $x_g = 2.2$ (or $E_g = 1.1$ eV) assuming a sun temperature of 6,000 K. The energy unit electon-volt 1 eV $= 1.60217662 \cdot 10^{-19}$ J is typically used in solid-state physics. A more accurate ultimate efficiency value can be computed by considering a sun temperature of 5,800 K or by integrating a measured solar spectrum such as the AM0 or AM1.5 instead of the Planck distribution [73].

3.6 Detailed Balance Limit

In order to compute a more accurate efficiency value, Schockley and Queisser [72] considered the various contributions to the generation and recombination of carriers in the solar cell, thus defining the detailed balance limit of the solar cell efficiency. There are five different contributions to the carrier generation and recombination, namely, (a) the generation of electron–hole pairs due to the absorption of the incident solar radiation with rate F_s; (b) radiative recombination of electron–hole pairs with rate $F_c(V)$; (c) removal of electron–hole pairs due to current generation from the solar cell with rate I/q and finally other nonradiative processes that result in (d) generation with rate $R(0)$; and (e) recombination with rate $R(V)$ of electron–hole pairs. The voltage V is the voltage between the p and n regions of the solar cell. In the steady-state condition, all the preceding processes are balanced, resulting in

$$I/q = F_s - F_c(V) + R(0) - R(V) \tag{3.11}$$

$$I = q\left[F_s - F_{c0}\right] + q\left[F_{c0} - F_c(V) + R(0) - R(V)\right], \tag{3.12}$$

where $F_{c0} = F_c(0)$. In the next paragraphs, we will consider the different processes separately in order to express the steady-state equation in a more practical form

$$\begin{aligned} I &= I_s + I_d = I_s - I_0\left[e^{(V/V_c)} - 1\right] \\ I_s &= q\left[F_s - F_{c0}\right] \\ I_d &= -q\left[F_{c0} - F_c(V) + R(0) - R(V)\right], \end{aligned} \tag{3.13}$$

where the different current components are identified as the short-circuit current I_s and the dark current I_d, with $I_0 = q\left[F_{c0} + R(0)\right]$, $V_c = kT_c/q$.

3.6.1 Generation of Electron–Hole Pairs Due to Solar Radiation

In order to compute the generation rate of hole–electron pairs, one first needs to compute the incident solar radiation. In computing the ultimate solar cell efficiency, it was assumed that all the power emitted from the sun reaches the solar cell, which was considered a black body with temperature T_s (3.9). A black body, however, radiates isotropically, and therefore only a fraction of the radiated power reaches the solar cell. In fact, the total number of photons per unit time reaching the surface of the solar cell is equal to the rate of generation of electron–hole pairs

$$F_s = f_w A Q_s, \tag{3.14}$$

where we assumed as before that the probability of excitation of a hole–electron pair by a photon with energy above the band gap E_g is one. A geometrical factor f_w has been introduced that depends on the solid angle ω_s subtended by the sun and the angle of incidence upon the solar cell.

The solid angle subtended by the sun is

$$\omega_s = \pi(D/L)^2/4 = 6.85 \cdot 10^{-5}\text{sr}, \tag{3.15}$$

where $D = 1.39 \cdot 10^6$ km is the diameter of the sun and $L = 149 \cdot 10^6$ km the distance of the sun to the earth and the solar cell.

Assuming a planar flat solar cell, the angle of incidence of the solar radiation is defined as the angle θ between the normal to the solar cell and the direction of the sun. In this case, the geometrical factor f_w becomes [72]

$$f_w = \omega_s \cdot \cos(\theta)/\pi. \tag{3.16}$$

In the case of normal incidence, the geometrical factor becomes $f_w = \omega/\pi = 2.18 \cdot 10^{-5}$.

Finally, the total incident power to the solar cell becomes

$$P_{in} = f_w A I_\lambda = f_w A \sigma T_s^4. \tag{3.17}$$

3.6.2 Radiative Recombination of Electron–Hole Pairs

In order to compute the recombination rate F_c of electron–hole pairs, Schockley and Queisser assumed that the solar cell is surrounded by a black body of temperature T_c. The recombination rate is proportional to the the number of electrons per unit area n in the conduction band and the number of holes per unit area p in the valence band, i.e., $F_c \propto np$.

In the equilibrium state, where no external perturbations such as the absorption of photons from the sun affect the solar cell, the rate of photons impinging to the solar cell by the surrounding black body is equal to the rate of photons radiated by the solar cell due to the recombination of electron–hole pairs. Consequently, no net current generation is observed. Similarly to the previous section, the rate of incident photons with energy above the bandgap is equal to

$$F_{c0} = 2AQ_c, \tag{3.18}$$

where $Q_c = Q(T_c, E_g, +\infty)$ the number of photons per unit time and area with energy above E_g computed by appropriately integrating Planck's distribution with temperature T_c. It is also assumed that each photon with such energy has a probability equal to 1 to generate an electron–hole pair. Furthermore, since the black body surrounds the solar cell, the total area of the cell is equal to $2A$ and the geometrical factor f_w is equal to one.

In the equilibrium state, the number of electrons per unit area n in the conduction band and the number of holes per unit area p in the valence band of the semiconductor are related by [61, 64]

$$np = n_i^2, \tag{3.19}$$

where $n_i \propto e^{-E_g/kT}$ is the intrinsic carrier density that depends on the bandgap energy E_g. When the solar cell is illuminated by an external light source such as the sun and electron–hole pairs are generated due to the absorption of photons, the solar cell is not in equilibrium and the carrier densities do not obey (3.19). However, when the disturbance is not very large and not too fast changing, the carrier densities reach a quasithermal equilibrium state that is modeled as a small perturbation of the original thermal equilibrium state [61, 64]. The carrier generation results in a chemical potential qV difference between the Fermi levels corresponding to the electrons and holes, where V is equal to the voltage generation across the semiconductor solar cell terminals. The electron and hole density product is then given

$$np = n_i^2 e^{V/V_c}. \tag{3.20}$$

Consequently, in the nonequilibrium case, the rate of radiative recombination becomes [72]

$$F_c(V) = F_{c0} e^{V/V_c}. \tag{3.21}$$

3.6.3 Nonradiative Generation and Recombination of Electron–Hole Pairs

The nonradiative generation and recombination rates of electron–hole pairs are defined as $R(0)$ and $R(V)$ respectively. Similarly to their radiative counterparts, they are also equal when the solar cell is in the equilibrium state, and they depend on the chemical potential qV in the nonequilibrium state as

$$R(V) = R(0)e^{V/V_c}. \tag{3.22}$$

Shockley and Queisser further assumed that the nonradiative processes are only a fraction of the radiative ones defined by the ratio $f_c = F_{c0}/(F_{c0} + R(0)$. Using 3.21 and 3.22, one obtains an expression for the dark current

$$\begin{aligned} I_d &= -q\left[F_{c0} - F_c(V) + R(0) - R(V)\right] \\ &= q\left[F_{c0} + R(0)\right]\left[e^{(V/V_c)} - 1\right] \\ &= I_0\left[e^{(V/V_c)} - 1\right], \end{aligned} \tag{3.23}$$

where

$$I_0 = q\left[F_{c0} + R(0)\right] = \frac{qF_{c0}}{f_c} = \frac{2qAQ_c}{f_c}. \tag{3.24}$$

3.6.4 The Short-Circuit Current and the Open-Circuit Voltage

The short-circuit current I_s and the open-circuit voltage V_{oc} are two important parameters characterizing the performance of a solar cell. The results of the previous section provide some insight into these two important parameters. One can find a relation between the two by setting $I = 0$ and $V = V_{oc}$ in (3.13) and solving for V_{oc} or I_s

$$\begin{aligned} V_{oc} &= V_c \ln\left(1 + \frac{I_s}{I_o}\right) \\ I_s &= I_o\left(e^{\frac{V_{oc}}{V_c}} - 1\right). \end{aligned} \tag{3.25}$$

The short-circuit current was defined from (3.12) and (3.13) to be

$$I_s = q\left[F_s - F_{c0}\right]. \tag{3.26}$$

We proceed by normalizing the expression of I_s using the saturation current I_o (3.24) to write

$$\frac{I_s}{I_o} = f_c\left[\frac{F_s}{F_{c0}} - 1\right]. \tag{3.27}$$

Finally, using (3.14) and (3.21), one obtains

$$\frac{I_s}{I_o} = f_c\left[\frac{f_w}{2}\frac{Q_s}{Q_c} - 1\right] \approx \frac{f_c f_w}{2}\frac{Q_s}{Q_c}. \tag{3.28}$$

The number of photons $Q(T, E_g, +\infty)$ with energy at least E_g emitted from a black body at temperature T_s has an exponential dependence on the band-gap energy E_g [72]

$$Q(T, E_g, +\infty) \approx A(T)e^{\frac{-E_g}{kT}}, \tag{3.29}$$

where $A(T)$ is a constant that depends on the temperature T. Using 3.29, one can obtain the dependence of the short-circuit current on the band-gap energy E_g

$$\frac{I_s}{I_o} \approx \frac{f_c f_w}{2} \frac{A(T_s)}{A(T_c)} e^{E_g/kT_c}. \tag{3.30}$$

Finally, using (3.25) and (3.30) one obtains the dependence of the open-circuit voltage on the band-gap energy

$$\begin{aligned} \frac{V_{oc}}{V_c} &= \ln\left[(1 - f_c) + \frac{f_c f_w}{2} \frac{Q_s}{Q_c}\right] \approx \ln\left[\frac{f_c f_w}{2} \frac{Q_s}{Q_c}\right] \\ \frac{V_{oc}}{V_c} &\approx \frac{E_g}{kT_c} + \ln\left[C(f_c, f_w, T_s, T_c)\right]. \end{aligned} \tag{3.31}$$

The constant C is smaller than 1, and consequently the open-circuit voltage is always smaller than the band-gap voltage E_g/q and becomes equal to it as temperature T_c tends to 0 K [72].

3.7 Circuit Model of Solar Cells

The current voltage characteristic of the solar cell is given by (3.13), which is repeated here for convenience.

$$I = I_s + I_d = I_s - I_0\left[e^{(V/V_c)} - 1\right]. \tag{3.32}$$

This is expressed in a circuit schematic as a current source in parallel with an ideal diode shown in Figure 3.3. When the equilibrium of the solar cell is perturbed by an external source of photons, additional carriers are generated in the cell. If the solar cell is connected to a load, then a current is generated. The value of this current corresponds to the short-circuit current I_s, and it is equal to the current due to the additional carriers qF_s minus a quantity qF_{c0} due to the recombination of carriers at thermal equilibrium [72], given in (3.13). Furthermore, as the external perturbation results in a quasithermal equilibrium, a voltage V is developed across the cell, which results in a current with opposite polarity called the dark current I_d of the solar cell. The dark current is equal to the current that flows across the solar cell if there is an external voltage V applied to its terminals in the dark (i.e., without any external light source applied). As one can immediately identify, the dark current expression is similar to that of an ideal diode.

The nonideal solar cell model also includes parasitic series R_s and parallel R_p resistances that represent thermal losses included in Figure 3.3. The series

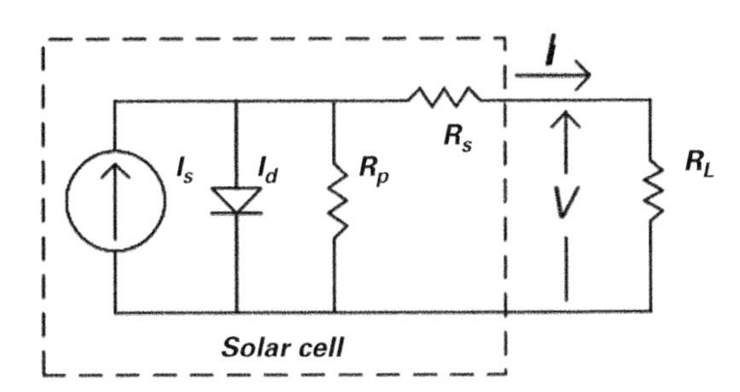

Figure 3.3 Circuit model of the solar cell.

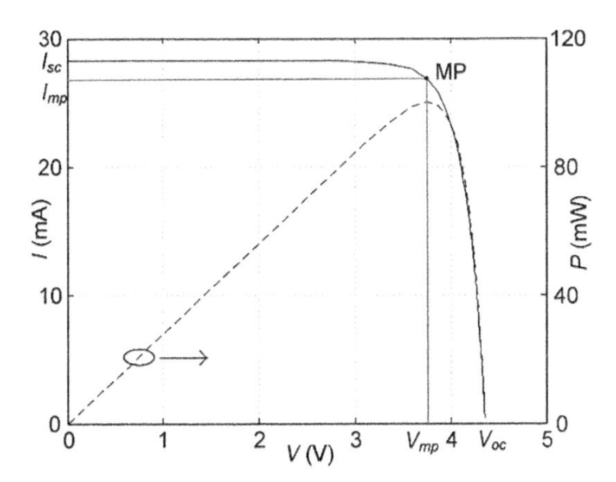

Figure 3.4 Current to voltage characteristic of an ideal solar cell.

resistance is typically due to the terminal contacts of the cell and the parallel resistance due to leakage current around the sides of the device [61]. The current voltage characteristic of a nonideal solar cell becomes

$$I = I_s - I_0 \left[e^{\frac{V+IR_s}{V_c}} - 1 \right] - \frac{V + IR_s}{R_p}. \tag{3.33}$$

3.8 The Detailed Balance Limit of Maximum Efficiency

The solar cell current to voltage I-V characteristic has the shape shown in Figure 3.4. Being a current source, the solar cell generates an approximately constant current for a wide range of voltage values across its terminals, which corresponds to different output loads connected to the solar cell. On one hand, as the load resistance becomes very small the solar cell current tends to its limiting value corresponding to the short-circuit current value I_s. On the other hand, as the load resistance takes large values the solar cell characteristic approaches the open-circuit voltage V_{oc}. The slope of the I-V characteristic near the I_s and

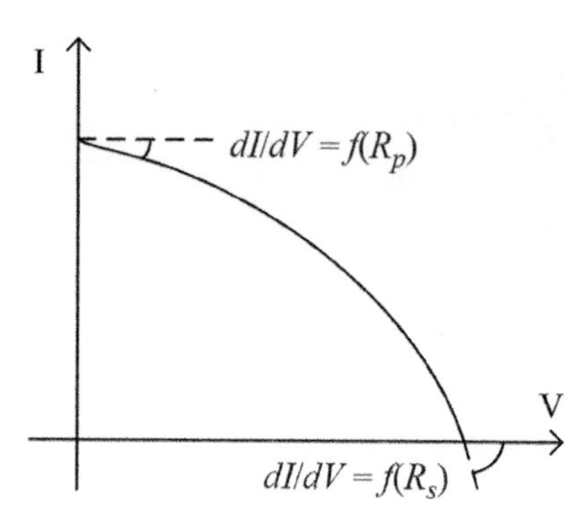

Figure 3.5 Effect of parasitic resistances on the current to voltage characteristic of a solar cell.

V_{oc} limiting values is dependent on the parasitic resistance R_s and R_p values respectively, as shown in Figure 3.5.

Due to the nonlinear nature of the I-V characteristic, there exists an optimum load value for a given solar irradiance or I_s, for which a maximum power P_m is delivered to the load as Figure 3.4 shows. This value, in effect, corresponds to the detailed balance limit of efficiency. The optimum voltage V_m can be obtained by taking the derivative of power versus the voltage and setting it equal to zero $dP/dV = d(IV)dV = 0$, which gives

$$\left(1 + \frac{V_m}{V_c}\right) e^{\frac{V_m}{V_c}} = e^{\frac{V_{oc}}{V_c}} = 1 + \frac{I_s}{I_o}. \tag{3.34}$$

The optimum voltage can be obtained by multiplying (3.34) with e in order to bring the left side to the form xe^x and using the definition of the Lambert function W to obtain

$$\frac{V_m}{V_c} = W_o\left(e(1 + I_s/I_o)\right) - 1 = W_o\left(e^{(1+V_{oc}/V_c)}\right) - 1, \tag{3.35}$$

where the principal branch W_o of the Lambert function is used because its argument takes positive values. The optimum current value I_m is then found solving (3.34) for e^{V_m/V_c} and substituting the obtained value in (3.13).

$$\frac{I_m}{I_o} = (1 + I_s/I_o)\left(\frac{V_m/V_c}{1 + V_m/V_c}\right) \tag{3.36}$$

Using (3.35) and (3.36), one can obtain an expression for the maximum output power from the solar cell $P_m = I_m V_m$,

$$P_m = (I_s + I_o)\frac{V_c\left(W_o\left(e(1 + I_s/I_o)\right) - 1\right)^2}{W_o\left(e(1 + I_s/I_o)\right)}. \tag{3.37}$$

The maximum power P_m is graphically given by a rectangle with sides equal to I_m and V_m. The values I_m and V_m are a fraction of I_s and V_{oc} respectively, which depends on (3.13) and (3.35). The fill factor FF expresses how close the product $I_m V_m$ is to the product $I_s V_{oc}$

$$FF = \frac{I_m V_m}{I_s V_{oc}}, \tag{3.38}$$

which represents a maximum possible power value for a given solar irradiance and solar cell and provides a measure of the quality of the solar cell. The detailed balance limit of efficiency is given

$$\eta = \frac{I_m V_m}{P_{in}} = \frac{FF I_s V_{oc}}{P_{in}}, \tag{3.39}$$

where the input power $P_{in} = f_w A I_\lambda = f_w A \sigma T_s^4$ has been derived in (3.17). Using (3.13), (3.14), (3.17), (3.21), (3.24), and (3.37), one obtains an expression for the maximum efficiency limit, which depends on T_s, T_c, f_w, f_c, and E_g,

$$\eta = \left[\frac{qQ_s}{I_\lambda}\right]\left[1 + \frac{2(1-f_c)}{f_c f_w}\frac{Q_c}{Q_s}\right]\frac{V_c\left[W_o\left(e(\frac{f_c f_w}{2}\frac{Q_s}{Q_c} + (1-f_c))\right) - 1\right]^2}{W_o\left(e(\frac{f_c f_w}{2}\frac{Q_s}{Q_c} + (1-f_c))\right)} \tag{3.40}$$

or, using the expression for the ultimate efficiency η_g (3.10),

$$\eta = \eta_g \eta_{sr} \tag{3.41}$$

with

$$\eta_{sr} = \left[\frac{qV_c}{E_g}\right]\left[1 + \frac{2(1-f_c)}{f_c f_w}\frac{Q_c}{Q_s}\right]\frac{\left[W_o\left(e(\frac{f_c f_w}{2}\frac{Q_s}{Q_c} + (1-f_c))\right) - 1\right]^2}{W_o\left(e(\frac{f_c f_w}{2}\frac{Q_s}{Q_c} + (1-f_c))\right)}, \tag{3.42}$$

where the effect of the geometrical considerations and of the radiative recombination in reducing the ultimate efficiency η_g has been included in the term η_{sr}. If one does not consider any nonradiative recombination processes $f_c = 1$, then the expression for the efficiency is simplified to

$$\eta = \eta_g \eta_{sr} = \eta_g \frac{qV_c}{E_g}\frac{\left[W_o\left(e(\frac{f_w}{2}\frac{Q_s}{Q_c})\right) - 1\right]^2}{W_o\left(e(\frac{f_w}{2}\frac{Q_s}{Q_c})\right)}. \tag{3.43}$$

Using the preceding theory, Schockley and Queisser [72] calculated the detailed balance limit of efficiency of solar cells for different parameter values, by deriving numerically the maximum current and voltage using the condition (3.34). The formulation obtained previously using the Lambert function provides an intuitive expression for the efficiency giving the same numerical result.

The efficiency for $T_s = 6{,}000$ K, $T_c = 300$ K, $f_w = 2.18 \cdot 10^{-5}$ and no nonradiative recombination $f_c = 1$ is plot in Figure 3.6. The numerical results showed that the maximum efficiency is approximately 30.5% for a band-gap energy $E_g = 1.314$ eV, whereas the maximum efficiency for Silicon with band-gap energy $E_g = 1.1$ eV is approximately 29.5% [72]. A more accurate efficiency limit can be obtained by considering a measured solar spectrum such as the AM0 or the AM1.5 instead of the Planck distribution [73].

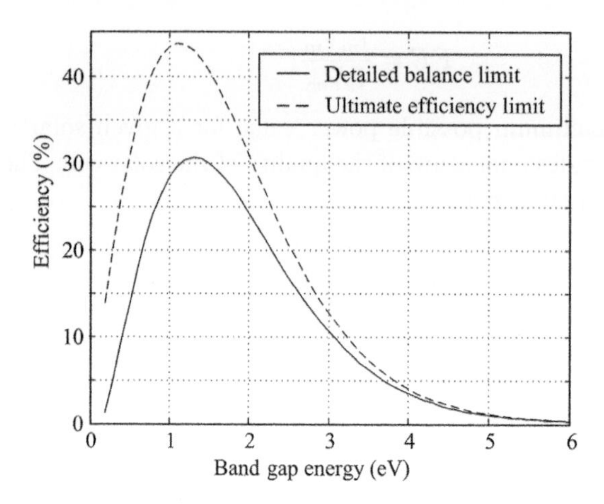

Figure 3.6 Ultimate efficiency and detailed balance efficiency limit versus bandgap energy E_g ($T_s = 6{,}000$ K, $T_c = 300$ K, $f_w = 2.18 \cdot 10^{-5}$ and no nonradiative recombination $f_c = 1$).

The four quantities I_s, V_{oc}, FF, and η are commonly used to compare the performance of different solar cells under various standard conditions such as $T_c = 25°$ C and AM1.5G solar irradiance. Comparison tables are published periodically in the literature [74]. Table 3.1 illustrates measured values of these parameters for a selective set of different solar cell technologies based on the data published in [74]. One can see that silicon solar cell performance is approaching the detailed balance efficiency limit, while the best efficiency to date is obtained by GaAs solar cells.

Table 3.1 Selected maximum measured solar cell performance parameters under AM1.5G solar irradiance at $T = 25$ ° C [74].

Technology	Efficiency (%)	V_{oc} (V)	I_s/A (mAcm^{-2})	FF (%)
Crystalline Si	26.7	0.738	42.65	84.9
Multicrystalline Si	22.3	0.6742	41.08	80.5
Amorphous Si	10.2	0.896	16.36	69.8
Thin film GaAs	29.1	1.1272	29.78	86.7
CIGS	22.9	0.744	38.77	79.5
CdTe	21.0	0.8759	30.25	79.4
Perovskite	20.9	1.125	24.92	74.5
Dye sensitized	11.9	0.744	22.47	71.2
Organic	11.2	0.78	19.30	74.2

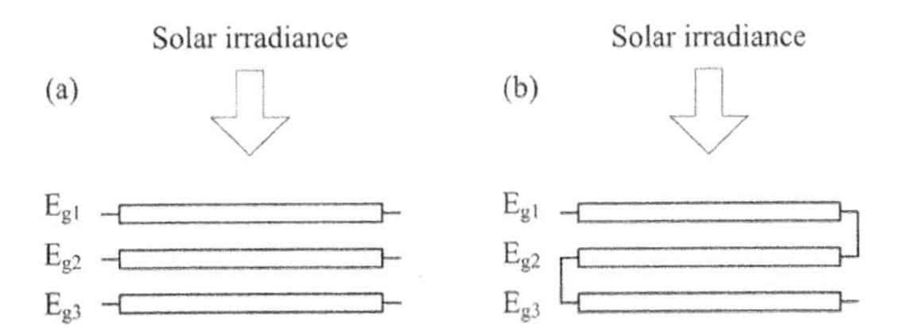

Figure 3.7 Architecture of a tandem solar cell: (a) unconstrained and (b) series connected.

3.9 Efficiency Limits for Tandem Solar Cells

It is possible to increase the overall efficiency of a solar cell by forming a stack of solar cells where each cell has an energy band gap that is optimum for converting to electrical energy the energy of the photons of a specific energy range (or equivalently frequency range) within the solar spectrum. These are called tandem solar cells [65].

We assume that a solar cell absorbs all photons with energy above its band gap and it is transparent to all photons with energy less than its band gap. This way we visualize a tandem cell comprising a stack of individual cells. Each solar cell in the tandem absorbs photons both from the solar radiation that reaches its surface and from the remaining solar cells of the tandem [75]. Due to the difference in temperatures between the sun and the solar cells, the amount of radiation from the remaining solar cells can be ignored when the number of cells in the tandem is small [75]. The top cell of the tandem, which the solar radiation reaches first, has the highest band-gap energy (E_{g1}), and the bad-gap energy of the subsequent cells in the stack is progressively reduced ($E_{g1} > E_{g2} > E_{g3} > \ldots$). The first (top) solar cell absorbs all photons with energy $E > E_{g1}$, the second cell absorbs photons with energy $E_{g2} < E < E_{g1}$, the third $E_{g3} < E < E_{g2}$, and so on. We may assume that the outputs of each solar cell are independent, resulting in what is called an unconstrained cell, or that they are connected in series [76]. The structure of the tandem cell is conceptually shown in Figure 3.7. The series connection imposes an additional constraint in the tandem cell in that the current through all solar cells has to be equal.

Therefore, both in the unconstraint and in the series connected tandem cell the band-gap energies of the individual cells may be optimized in order to find the values that lead to a maximum obtained efficiency. In order to compute the optimum band-gap energies, it is possible to consider either the ultimate solar cell efficiency (Section 3.5) or the detailed balance limit of efficiency (Section 3.8) and model the solar radiation either as a blackbody or using measured spectra such as the AM1.5G spectrum.

Based on the preceding, in order to compute the solar cell efficiency of the tandem cell we will need to integrate the Plank distribution corresponding to a blackbody with temperature T between two energy limits E_1 and E_2, defined

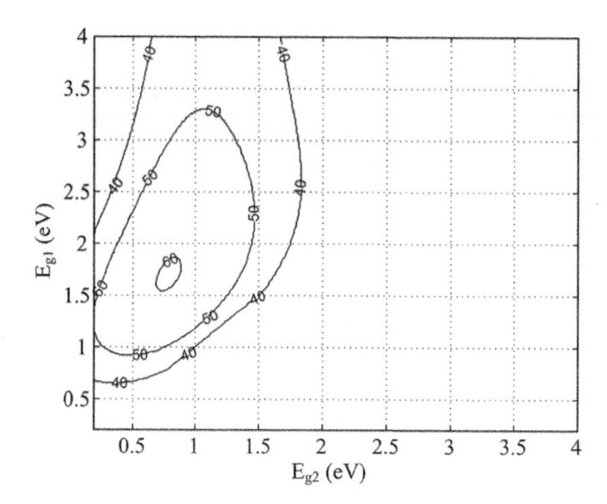

Figure 3.8 Ultimate efficiency limit for a tandem solar module comprising two cells with different band-gap energy.

in Equation (3.6). Using the efficiency formulations derived in Sections (3.5) and (3.8), we have computed efficiency contours for a two band-gap tandem cell modeling the solar radiation as a blackbody with $T_s = 6{,}000$ K. The ultimate efficiency for each solar cell is computed based on Equation (3.10), where the integral of the Planck distribution Q is computed using the appropriate energy limits for each solar cell in the tandem. The ultimate efficiency of the tandem cell becomes

$$\eta_g = \frac{E_{g1}Q_{s1} + E_{g2}Q_{s2}}{I_\lambda} \tag{3.44}$$

with $Q_{s1} = Q(T_s, E_{g1}, +\infty)$ and $Q_{s2} = Q(T_s, E_{g2}, E_{g1})$. The results corresponding to the ultimate efficiency limit for an unconstrained tandem are shown in Figure 3.8. The optimum band-gap energies corresponding are found to be approximately 0.78 eV and 1.7 eV.

The detailed efficiency limit was also computed in Figure 3.9 assuming $T_c =$ 300 K, $f_w = 2.18 \cdot 10^{-5}$ and no radiative recombination. In this case, the efficiency is found based on (3.40) with $f_c = 1$. The detailed efficiency limit of the tandem cell takes the form

$$\eta = \frac{I_{m1}V_{m1} + I_{m2}V_{m2}}{P_{in}}, \tag{3.45}$$

where $P_{m1} = I_{m1}V_{m1}$ and $P_{m2} = I_{m2}V_{m2}$ are the maximum power values from each cell in the tandem and the input power $P_{in} = f_w A I_\lambda = f_w A \sigma T_s^4$ has been defined in (3.17). The maximum voltage and current of a solar cell have been defined in (3.35) and (3.36) respectively and for $f_c = 1$ using (3.28) and (3.24), they become

$$V_m = V_c \left[W_o \left(e \left(\frac{f_w Q_s}{2Q_c} \right) \right) - 1 \right] \tag{3.46}$$

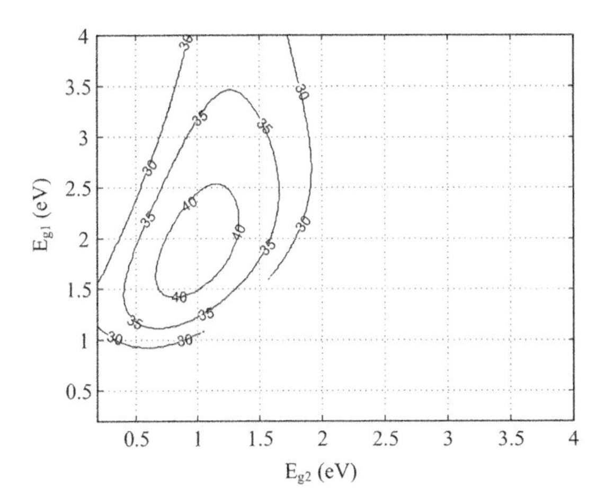

Figure 3.9 Detailed balance efficiency limit for a tandem solar module comprising two cells with different band-gap energy.

$$I_m = qAf_wQ_s \frac{\left[W_o\left(e\left(\frac{f_wQ_s}{2Q_c}\right)\right) - 1\right]}{W_o\left(e\left(\frac{f_wQ_s}{2Q_c}\right)\right)}. \tag{3.47}$$

For the two solar cells of the tandem, the various integrals of the Planck distribution become $Q_{c1} = Q(T_c, E_{g1}, +\infty)$, $Q_{c2} = Q(T_c, E_{g2}, +\infty)$ and, as before, $Q_{s1} = Q(T_s, E_{g1}, +\infty)$, $Q_{s2} = Q(T_s, E_{g2}, E_{g1})$. This way, the detailed efficiency limit of the tandem solar cell becomes

$$\eta = \frac{qV_c}{I_\lambda}\left[\frac{Q_{s1}\left[W_o\left(e(\frac{f_w}{2}\frac{Q_{s1}}{Q_c1})\right) - 1\right]^2}{W_o\left(e(\frac{f_w}{2}\frac{Q_{s1}}{Q_c1})\right)} + \frac{Q_{s2}\left[W_o\left(e(\frac{f_w}{2}\frac{Q_{s2}}{Q_c2})\right) - 1\right]^2}{W_o\left(e(\frac{f_w}{2}\frac{Q_{s1}}{Q_c2})\right)}\right]. \tag{3.48}$$

Finally, Figure 3.10 shows the computed detailed efficiency limit in the case of a series constrained tandem cell. In this case, we must limit the maximum current of the tandem cell to the value corresponding to the smaller of the two cells and then compute the corresponding maximum voltage values. This is done, for example, defining an intermediate parameter X as follows:

$$X = \min\left\{Q_{s1}\frac{\left[W_o\left(e\left(\frac{f_wQ_{s1}}{2Q_{c1}}\right)\right) - 1\right]}{W_o\left(e\left(\frac{f_wQ_{s1}}{2Q_{c1}}\right)\right)}, Q_{s2}\frac{\left[W_o\left(e\left(\frac{f_wQ_{s2}}{2Q_{c2}}\right)\right) - 1\right]}{W_o\left(e\left(\frac{f_wQ_{s2}}{2Q_{c2}}\right)\right)}\right\}. \tag{3.49}$$

The series current is then defined with the help of X as follows:

$$I_m = qAf_wX. \tag{3.50}$$

Having found the current through the solar cell I_m, we can solve (3.13) to compute the solar cell voltage,

$$V_m = V_c \ln\left[1 + \frac{I_s - I_m}{I_o}\right], \tag{3.51}$$

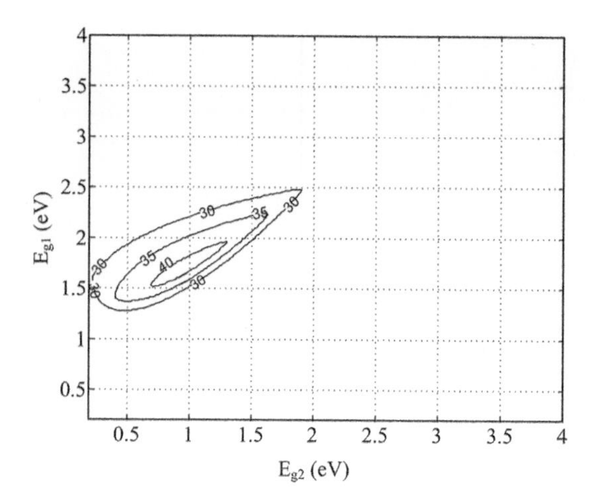

Figure 3.10 Detailed balance efficiency limit for a tandem solar module comprising two series connected cells with different band-gap energy.

where $I_o = 2qAQ_c$ is found from in (3.24) using $f_c = 1$. Additionally using (3.28), this becomes for each of the two solar cell voltages

$$V_{m1} = V_c \ln\left[\frac{f_w}{2} \cdot \frac{Q_{s1} - X}{Q_{c1}}\right] \tag{3.52}$$

and

$$V_{m2} = V_c \ln\left[\frac{f_w}{2} \cdot \frac{Q_{s2} - X}{Q_{c2}}\right]. \tag{3.53}$$

Finally, the detailed efficiency of the tandem cell becomes

$$\eta = \frac{I_m(V_{m1} + V_{m2})}{f_w A I_\lambda} = \frac{qX(V_{m1} + V_{m2})}{I_\lambda}. \tag{3.54}$$

The maximum efficiency of the unconstrained tandem cell with two cells was found to be 42.3% and for the series connected tandem 41.7%. This value represents a substantial increase compared to the maximum efficiency of a single cell of 30.5% shown in Figure 3.6. This formulation can easily be extended to a tandem cell with more than two cells.

3.10 Solar Antennas and Rectennas

Solar antennas integrate solar panels with antennas and have been originally proposed for satellite applications [77]. As a result, antenna and solar panel arrays have been developed [78, 79]. The immediate advantage of integrating solar cells and antennas is that of reducing size and consequently cost. The recent interest in extending the energy autonomy of wireless sensor nodes in the context of the Internet of Things (IoT) has resulted in design efforts focusing on low-profile solar antennas implemented in rigid substrates [80], transparent substrates [81], and also flexible substrates including paper, plastic polyethylene terephthalate (PET) [82], and textiles [83].

In the design of solar antennas, different types of printed antennas can be used, including patch or dipole/monopole antennas on the one hand and slot type antennas [79] on the other hand. Furthermore, the design process takes advantage of the fact that solar cells typically include a conductive (mirror) layer that can be integrated on top of the conductive surface of the antenna, or alternatively play the role of the conductive surface of the antenna itself if the layer conductivity is sufficiently high [79]. The design process consists of identifying through electromagnetic simulation the optimal placement of the solar cell in order not to disturb the radiation and impedance properties of the antenna. Another design challenge is that of identifying the proper placement of the wiring associated with the terminals of the solar cell and embedding it in the radiating structure. In this case, it is often common to connect one of the solar cell terminals (e.g., the dc ground) directly onto one of the conducting surfaces of the antenna, while the second terminal must be carefully wired in order not to disturb the radiating fields of the antenna.

An example of an ultrawideband monopole on PET substrate with a flexible solar cell on top is shown in Figure 3.11 [84]. The positive terminal of the solar cell is connected to the monopole conductor, while the negative terminal of the solar cell is connected to the ground conductor of the monopole through a conductive line near the top of the monopole, which has been shown in simulation to have a minimum effect in distorting the input impedance and the radiation pattern of the monopole antenna. The input s-parameters of the antenna with and without the solar cell are shown in Figure 3.11b, where we can see that the solar cell affects very little the input impedance of the antenna. Furthermore, measured radiation patterns of the solar antenna are shown in Figure 3.12 [85]. One can see that the presence of the solar cell has little effect on the antenna maximum gain and copolarization radiation pattern, and it results in an increase in the cross-polarization radiation pattern.

An example of a slot antenna integrating cells on top of the conducting areas surrounding the slot is shown in Figure 3.13 [84]. The antenna is a cavity backed slot antenna, where the cavity is formed by metalized vias in the substrate implementing an antenna in substrate integrated circuit technology.

An example of a textile solar antenna is shown in Figure 3.14 [83]. A quarter wavelength shorted patch antenna is used as the basis of the solar antenna, and the solar cell wiring is taken from the shorted side of the patch, thus having a minimum effect on the antenna performance.

Table 3.2 includes selected recent solar antenna designs in the literature and summarizes their properties.

Finally, one interesting application is that of integrating solar cells with RFID tag antennas in order to provide an additional source of power to the tag other than the RF power and consequently increase their operating range. The idea

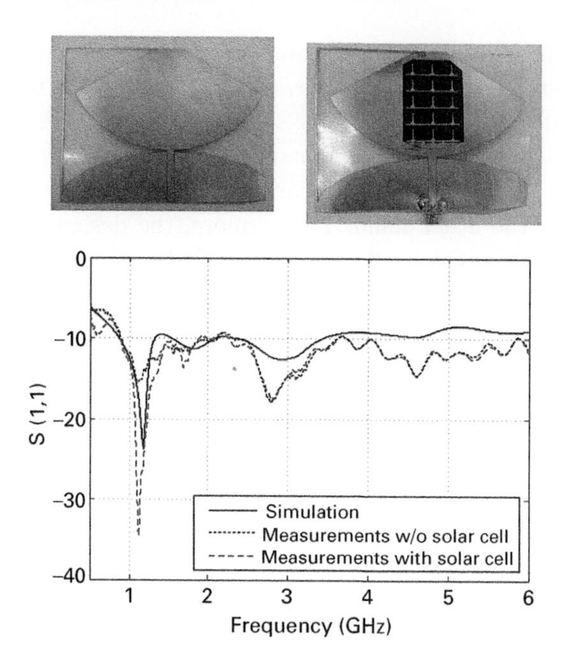

Figure 3.11 Flexible solar cell integrated on top of an ultrawideband monopole on a PET substrate demonstrated in [84]: photo of the antenna with and without the solar cell and input s-parameters. ©2014 IEEE. Reprinted, with permission from [84]

Table 3.2 Selected solar antenna publications.

Reference	Antenna type	Frequency
[77]	Patch antenna, linear polarization	2.25 GHz
[79]	Slot antenna, Ultralam substrate	3.87 GHz
[80]	UWB monopole antenna, FR4 substrate	3.1 GHz–10.6 GHz
[83]	Shorted patch antenna, textile substrate	915 MHz
[82]	UWB monopole, PET substrate	0.85 GHz–6 GHz
[81]	Patch antenna, transparent conductor AgHT-4	3.4 GHz–3.8 GHz
[86]	Slot antenna array underneath solar cell	2.4 GHz

was initially proposed in [87], where a printed monopole antenna with a solar cell on top of the monopole ground and an RFID tag was envisioned. The challenge is how to supply the RFID tag IC with the dc power from the solar cell. While there exist tags with an external input port separate from the RF port that accepts dc power, tag ICs typically only have two RF input ports. In [88], solar cells were integrated on RFID tag dipole arms (Figure 3.15) and the dc signal from the solar cells was converted to an RF signal by powering an RF oscillator placed next to the tag IC. The RF signal from the oscillator was fed to the input RF ports of the tag IC, therefore enabling the topology to be compatible with common type RFID tag ICs. The solar RFID tag is shown in Figure 3.15.

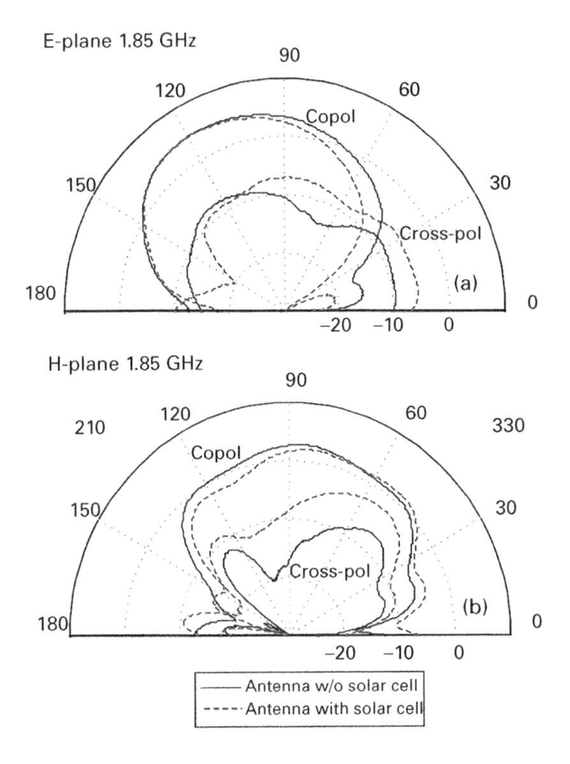

Figure 3.12 Measured (a) E-plane and (b) H-plane radiation patterns of the solar monopole antenna shown in Figure 3.11a at 1.85 GHz [85].

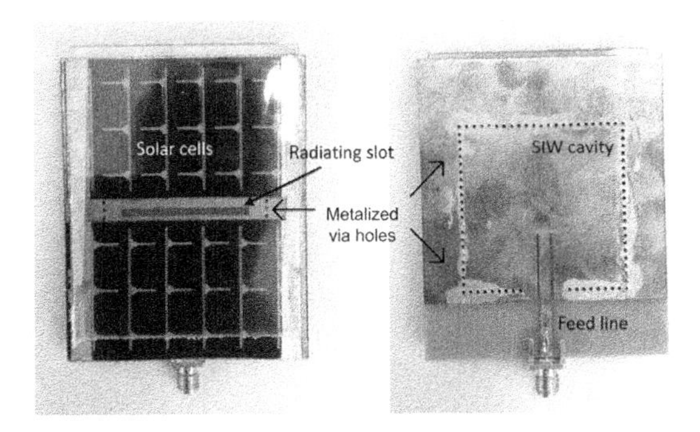

Figure 3.13 Cavity-backed substrate-integrated waveguide slot antenna with solar cells. ©2014 IEEE. Reprinted, with permission from [84]

3.11 Problems and Questions

1. What is the physical phenomenon that the operation of solar cells is based on, when was it discovered, and by who?
2. What is the measure 1 sun?
3. What is the AM1.5G standard?

Figure 3.14 Solar patch antenna on textile substrate courtesy of Prof. Hendrik Rogier, Ghent University [83].

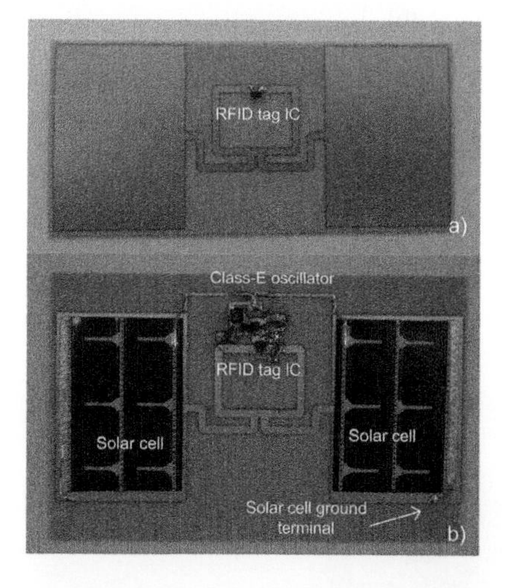

Figure 3.15 Solar power harvesting assisted RFID tag: a) original tag and b) modified tag with solar cells and oscillator circuit. ©2012 IEEE. Reprinted, with permission from [88]

4. What is the ultimate solar cell efficiency?
5. What is the detailed balance limit?
6. Compute the maximum value of ultimate solar cell efficiency and the corresponding band-gap value corresponding to the AM0 and AM1.5G spectra.
7. Name four characterizing parameters of solar cells. What is the fill factor of a solar cell?
8. What is the maximum solar cell efficiency based on the detailed balance limit for a silicon solar cell ($E_g = 1.1$ eV) and for a GaAs solar cell ($E_g = 1.424$ eV),

assuming $T_s = 6{,}000$ K, $T_c = 300$, $f_w = 2.18 \cdot 10^{-5}$, and no radiative recombination.

9. Compute the ultimate efficiency contours of tandem solar cells with three band gaps. What are the optimum band-gap energies and what is the maximum theoretical efficiency?

4 Kinetic Energy Harvesting

4.1 Introduction

Kinetic energy harvesters convert energy from mechanical movements to electrical. A large number of applications can be classified under this category as there are many different types of mechanical movements such as vibrations, displacements, and forces or pressure and consequently different types of transducing mechanisms.

One may identify applications related to buildings and other construction projects such as bridges, exploring vibrations originating from a plurality of sources such as nearby or ongoing car automotive or train traffic; human movement; wind; heating; ventilating, and air conditioning (HVAC) air currents; and water and other fluid movements [63]. Vibration energy can be harvested in moving vehicles such as cars, trains, ships, and airplanes by properly installing transducers in sensitive places of the vehicle, such as the wheels of the car.

One may also distinguish between large-scale projects such as wind farms or hydroelectric energy plants and microenergy harvesting systems. Finally, one should mention kinetic energy harvesting applications involving human body movement [89], such as automatic watches harvesting energy from human arm movement and more recent applications harvesting energy from walking (or running) activities [10].

The chapter begins with a description of the three different transducing types associated with kinetic energy harvesting and then continues with a mathematical model and theoretical expressions for the maximum harvester energy and efficiency. A comparison of the performance of the the transducer types follows with a brief discussion of figures of merit. Finally, some selected examples of kinetic energy harvesters are provided.

4.2 Transducer Types

There are three transducing mechanisms that are used to convert kinetic energy to electrical energy: (a) electrostatic (capacitive), (b) electromagnetic (inductive), and (c) piezoelectric. The principle of operation and physics behind each mechanism is described in the following subsections.

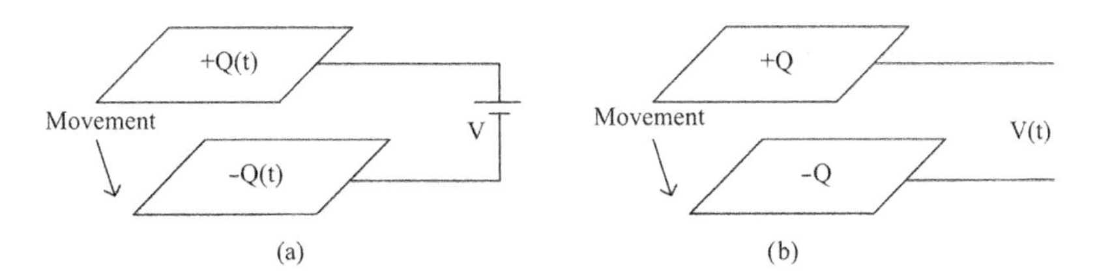

Figure 4.1 Electrostatic energy harvesting transducer: (a) fixed voltage and (b) fixed charge.

4.2.1 Electrostatic Transducers

Electrostatic transducers produce electrical energy by varying (a) the stored charge in a variable capacitor or (b) the voltage of a variable capacitor (Figure 4.1) [90]. The stored energy in a capacitor C is given by

$$E = \frac{1}{2}QV = \frac{1}{2}CV^2 = \frac{1}{2}\frac{Q^2}{C}, \tag{4.1}$$

where $Q = CV$ is the stored charge and V the voltage across its plates.

In the first method, in a first step, a variable capacitor is charged to a voltage V_{max} at a maximum capacitance C_{max}. In a second step, the capacitance is reduced to C_{min} employing, for example, mechanical movement of the capacitor plates, while the voltage across its plates is maintained at V_{max}. The produced energy by this procedure is

$$\Delta E = \frac{1}{2}\left(C_{max} - C_{min}\right)V_{max}^2. \tag{4.2}$$

The disadvantage of this method is that a voltage generator is required in order to maintain the voltage across the capacitor plates constant.

In the second method, the capacitor is first charged to a certain charge at the maximum capacitance, which creates a voltage $Q = C_{max}V_1$ across its terminals. The capacitance is then decreased to a value C_{min}, using for example mechanical means, while the capacitor terminals are kept as an open circuit in order to maintain its charge Q. As a result, the voltage across its plates is increased to $V_{max} = Q/C_{min}$ and electrical energy is produced:

$$\Delta E = \frac{1}{2}Q\left(V_{max} - V_{min}\right) = \frac{1}{2}\left(C_{max} - C_{min}\right)V_{max}V_1. \tag{4.3}$$

One can verify by comparing (4.2) and (4.3) that the amount of energy that is produced by the method of constant charge is less than the one produced by the architecture of constant voltage by a factor of V_{max}/V_1. However, the method of constant charge does not require any voltage generator circuitry, and consequently it is more easily implemented. Both architectures, however, require an initial charging phase of the capacitor before energy can be harvested. Nonetheless, a very attractive property of electrostatic transducers is that they can easily be integrated with other electronic circuitry, as microelectromechanical

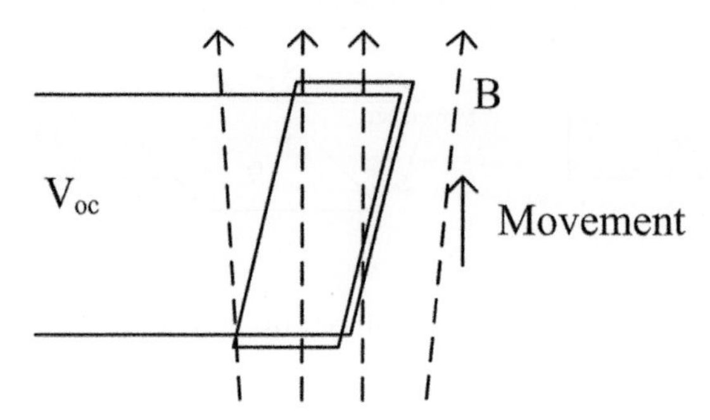

Figure 4.2 Electromagnetic energy harvesting transducer.

systems (MEMS) can be used to fabricate variable capacitor-based electrostatic transducers [63, 91].

4.2.2 Electromagnetic Transducers

Electromagnetic transducers are based on Faraday's law of induction. A schematic of such a transducer is shown in Figure 4.2. Specifically, an electromotive force, a voltage difference, is produced between the edges of a conductor when it moves inside a magnetic field. Typically, a coil is used as the conducting material and a permanent magnet generates the required magnetic field. Mechanical movement of the coil relative to the magnet results in a time-varying magnetic flux through the coil and subsequently an open-circuit voltage V_{oc}, which is equal to

$$V_{oc} = -N\frac{d\Phi}{dt}, \tag{4.4}$$

where N is the number of turns of the coil and Φ the magnetic field flux. Well-known examples are Faradays disc [92], a rotating coil inside a uniform static magnetic field, and Tesla's dynamo [93].

If one considers a coil that is moving along a path $x(t)$ inside a magnetic field B, (4.4) becomes

$$V_{oc} = -N\frac{d\Phi}{dx}\frac{dx}{dt} = D\frac{dx}{dt}. \tag{4.5}$$

In the case of a coil moving along the direction x with its area A oriented perpendicularly to a magnetic field with a magnitude gradient dB/dx, as shown in Figure 4.2, the open-circuit voltage becomes

$$V_{oc} = -NA\frac{dB}{dx}\frac{dx}{dt}. \tag{4.6}$$

An example of such a topology was used in [94], where the authors estimated that about 400 μW could be generated on average. Most of the rotating-type generators found today, from bicycle dynamos to large-scale power generation plants such as hydroelectric installations and wind farms, are electromagnetic transducers [94].

4.2.3 Piezoelectric Transducers

The third type of mechanical to electric energy transducers is piezoelectric transducers. Piezoelectric materials produce an electric field when they are deformed. Conversely, when an electric field is applied to a piezoelectric material, it changes its form. The piezoelectric phenomenon is attributed to the presence of electric dipole moments within the material [63]. Application of pressure causes the microscopic dipole moments to align and results in a macroscopic polarization vector and a voltage difference along the material. The constitutive relations of a piezoelectric material are as follows [95, 96]:

$$\begin{aligned} \boldsymbol{S} &= \sigma \boldsymbol{T} + d^T \boldsymbol{E} \\ \boldsymbol{D} &= d\boldsymbol{T} + \epsilon \boldsymbol{E}, \end{aligned} \tag{4.7}$$

where $\boldsymbol{S}$ is the strain, $\boldsymbol{T}$ is the stress, $\boldsymbol{E}$ the electric field, and $\boldsymbol{D}$ the electric displacement. Bold type is used to indicate vector quantities. Furthermore, σ and ϵ are the compliance and permittivity tensors respectively. The piezoelectric strain is also a tensor quantity indicated by d. The superscript $()^T$ is used to indicate a matrix transpose. In absence of piezoelectric strain, the second equation becomes the standard constitutive equation between displacement $\boldsymbol{D}$ and electric field $\boldsymbol{E}$.

Piezoelectric materials are anisotropic materials, and as such they have a different behavior depending on the direction of the stress $\boldsymbol{T}$ and the electric field $\boldsymbol{E}$. Specifications of piezoelectric materials typically provide a piezoelectric strain constant d using two subscript indices, the first one referring to the direction of the electric field and the second to the direction of the stress. The three axes x, y and z are indicated using the indices 1, 2, and 3 respectively. As an example, a piezoelectric strain constant d_{33} refers to a piezoelectric system topology where both the electric field and the stress are applied along the z-direction that corresponds to the thickness of the piezoelectric material, as shown in Figure 4.3 [63]. Commonly used piezoelectric materials are ferroelectric perovskites and piezoelectric polymer materials [63, 96]. A characteristic representative of perovskite ceramic materials is lead zirconatetitanate Pb(Zr,Ti)O_3, (PZT), whereas a commonly used piezoelectric polymer is polyvinylidene fluoride (PVDF).

4.3 Modeling Vibration Energy Harvesting Systems

Williams and Yates have proposed a model for electric generators harvesting energy from a mechanical vibration motion that is widely used [97]. The model consists of a mass m that is attached to a spring of stiffness k from a frame that is excited by an external source y as shown in Figure 4.4. As a result of the external excitation y, the mass m is displaced relative to the frame by z. Losses in the transfer of kinetic energy from the external source y to the mass m are indicated by the damping coefficient $d_T = d_m + d_e$, which includes

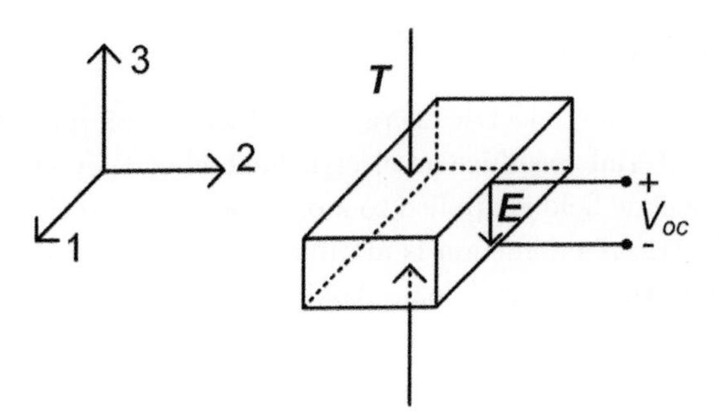

Figure 4.3 Piezoelectric strain constant d_{33} example system.

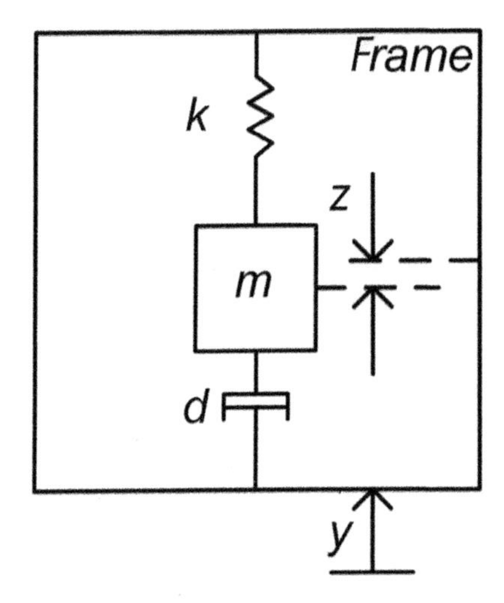

Figure 4.4 Model of a electric energy generator based on kinetic energy harvesting [97].

both mechanical parasitic losses d_m as well as losses d_e corresponding to energy conversion from mechanical to electrical. The system dynamics are described by the second-order differential equation

$$m\ddot{z} + d_T\dot{z} + kz = -m\ddot{y}. \tag{4.8}$$

The system differential equation can be expressed in a normalized form

$$\ddot{z} + 2\zeta_T\omega_n\dot{z} + \omega_n^2 z = -\ddot{y}, \tag{4.9}$$

where ω_n is the natural frequency of oscillation of the system and

$$\zeta_T = \frac{d}{2m\omega_n} \tag{4.10}$$

is the damping factor, which comprises two contributions, ζ_m corresponding to parasitic mechanical losses d_m and ζ_e corresponding to the mechanical-to-electrical conversion losses d_e. One obtains the steady-state response of the system by taking the Fourier transform of (4.8) or (4.9),

$$Z(\omega) = \frac{\left(\frac{\omega}{\omega_n}\right)^2}{1 - \left(\frac{\omega}{\omega_n}\right)^2 + j2\zeta_T\left(\frac{\omega}{\omega_n}\right)} Y(\omega) = A(\omega)Y(\omega). \tag{4.11}$$

Assuming a sinusoidal excitation $y(t) = Y_o \sin(\omega t)$ and taking the inverse Fourier transform of (4.11), one obtains the time domain displacement $z(t)$ of the mass m as

$$z(t) = \frac{\left(\frac{\omega}{\omega_n}\right)^2 Y_o}{\sqrt{\left[1 - \left(\frac{\omega}{\omega_n}\right)^2\right]^2 + \left(2\zeta_T \frac{\omega}{\omega_n}\right)^2}} \sin(\omega t + \phi) \tag{4.12}$$

with

$$\phi = -\tan^{-1}\left(\frac{2\zeta_T \frac{\omega}{\omega_n}}{1 - \left(\frac{\omega}{\omega_n}\right)^2}\right). \tag{4.13}$$

The instantaneous electrical power P_e that is generated is the power dissipated due to d_e, which is equal to the force $F_e = d_e \dot{z} = 2\zeta_e m\omega_n \dot{z}$ corresponding to the damping coefficient d_e, multiplied by the velocity $\dot{z}$, i.e., $P_e = F_e \dot{z}$. The average generated electrical power is therefore given by

$$P_e = d_e < \dot{z}^2 > = \; 2\zeta_e m\omega_n < \dot{z}^2 >, \tag{4.14}$$

where $<>$ denotes a time average. Taking the time derivative of the steady-state solution (4.12) in order to find $\dot{z}$ and applying it in the expression for the average generated power (4.14), one calculates the average electrical power generated as

$$P_e = \frac{\zeta_e m\omega_n \left(\frac{\omega}{\omega_n}\right)^4 \omega^2 Y_o^2}{\left[1 - \left(\frac{\omega}{\omega_n}\right)^2\right]^2 + \left(2\zeta_T \frac{\omega}{\omega_n}\right)^2} = \frac{\zeta_e m \left(\frac{A^2\omega^2}{\omega_n^3}\right)}{\left[1 - \left(\frac{\omega}{\omega_n}\right)^2\right]^2 + \left(2\zeta_T \frac{\omega}{\omega_n}\right)^2}, \tag{4.15}$$

where $A = \omega^2 Y_o$ is the acceleration of the external excitation y. Maximum power conversion occurs when the excitation y has the same frequency as the natural frequency of the system $\omega = \omega_n$, resulting in

$$P_{em} = \frac{m\zeta_e \omega_n^3 Y_o^2}{4\zeta_T^2} = \frac{m\zeta_e A^2}{4\zeta_T^2 \omega_n}. \tag{4.16}$$

The generated power is strongly dependent on how close the excitation frequency ω is to the system natural frequency ω_n. This is visualized next by expressing the average generated power relative to its maximum value as

$$\frac{P_e}{P_{em}} = \frac{4\zeta_T^2 \left(\frac{\omega}{\omega_n}\right)^2}{\left[1 - \left(\frac{\omega}{\omega_n}\right)^2\right]^2 + 4\zeta_T^2 \left(\frac{\omega}{\omega_n}\right)^2}. \tag{4.17}$$

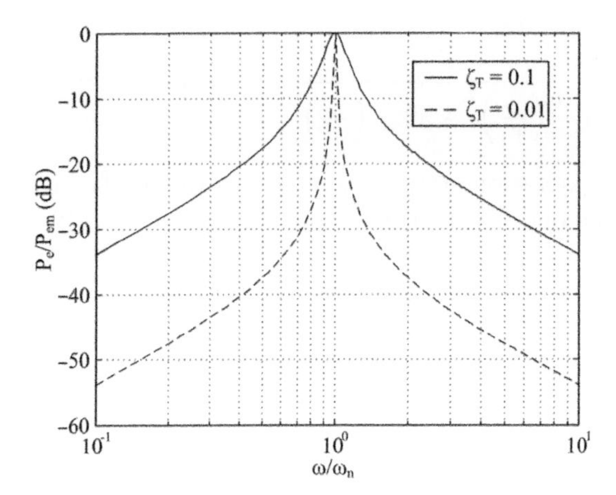

Figure 4.5 Dependence of the generated power on frequency for different damping coefficients.

The variation of the generated power with respect to the frequency of the external excitation relative to the resonance frequency of the system is shown in Figure 4.5 for $\zeta_T = 0.1$ and $\zeta_T = 0.01$.

The strong dependence on the natural resonance frequency of the system is one of the main drawbacks of vibration to electric transducers as it makes their design very application dependent. One should first accurately determine by simulation or measurement the frequency of excitation corresponding to the specific application scenario and subsequently design a generator with the same natural frequency of oscillation. A great challenge in the vibration to electrical transducer design is that of maximizing harvested energy by controlling the natural frequency of the transducers. There are several different possibilities proposed in the literature that include transducers with mechanically or electrically tunable resonant frequencies and self-tuning or adaptive tuning capability on one hand, and wideband and multiband arrays of generators on the other hand [63, 98].

One may further maximize P_e by taking its derivative versus the damping factor ζ_e and setting it equal to zero in order to solve for the optimum damping factor $\zeta_e = \zeta_{em}$. A straightforward calculation gives that the optimum value of the damping factor ζ_{em} is equal to half the total damping factor ζ_T or, in other words, when the electrical and parasitic mechanical damping factors are equal, $\zeta_{em} = \zeta_m = \zeta_T/2$, which gives

$$P_{emo} = \frac{m\omega^3 Y_o^2}{16\zeta_{em}} = \frac{mA^2}{16\zeta_{em}\omega_n}. \tag{4.18}$$

Therefore, the maximum generated power is proportional to the square of the external signal acceleration amplitude A, proportional to the mass m of the system and inversely proportional to the natural resonance frequency ω_n of the system and the damping factor ζ_{em}. In general, in order to maximize the electrical output power of the system, the mass of the system should be maximized and the system should be operated in its lower resonance frequency [99].

Finally, one should emphasize the generality of the model by Williams and Yates, which can be used to model any kind of vibration to electric energy transducer, electrostatic, electromagnetic, or piezoelectric. One limitation of the model is that the conversion to electric energy is taken into account by considering a linear term in the system differential equation through a damping coefficient, which may not always be accurate [91]. One of the most important tasks of the designer becomes that of computing and optimizing the value of the damping factor for the system under consideration, which is typically evaluated using numerical techniques such as a finite element method simulator [63].

4.4 Vibration Sources

The theoretical analysis has shown that the maximum generated electrical power depends on the acceleration A and the frequency $\omega_n = 2\pi f_n$ of the external vibration sources. A fundamental task of the designer of a kinetic energy harvesting system is to determine these parameters for the application under consideration. Such an analysis was carried out in [100], where several external vibration source scenarios were studied and the acceleration and vibration frequencies were evaluated. Table 4.1 includes a selection of the results published in [100]. One can see that the source frequencies are in the order of a few tens to a few 100 Hz, whereas the acceleration values range from 0.1 ms^{-2} to 3 ms^{-2}.

Table 4.1 Measured frequencies and acceleration amplitude of various vibration sources [100].

Application, vibration source	Vibration frequency (Hz)	Acceleration amplitude (ms^{-2})
Door frame just as door closes	125	3
Clothes dryer	121	3.5
Washing machine	109	0.5
HVAC vents in office building	60	0.2–1.5
Refrigerator	240	0.1
Small microwave oven	121	2.25
External windows (2 ft × 3 ft) next to a busy street	100	0.7

4.5 Comparison of Different Kinetic Energy Harvesters

An immediate question that comes to mind is how do the different types of vibration harvesters compare with each other. In order to address this question, a linear, two-port model for a general kinetic energy transducer was considered in [100]. The performance of the transducer is determined by two parameters, the coupling coefficient κ and the transmission coefficient λ. The transmission coefficient λ is equal to the efficiency of the transducer, the ratio of the energy delivered to the load to the average input energy to the transducer. The coupling coefficient κ instead is equal to the energy stored within the transducer over the average input energy. The maximum transmission coefficient λ_m was found to be equal to [100]

$$\lambda_m = \frac{\kappa^2}{4 - 2\kappa^2}. \tag{4.19}$$

Assuming a harmonic external excitation with frequency ω and input energy U_i, the maximum power delivered to the load is

$$P_m = \lambda_m \omega U_i. \tag{4.20}$$

The coupling coefficient κ was computed for different types of kinetic energy transducers in [100], which allows one to compare the different transducers. Specifically, the coupling factor of an electromagnetic generator was evaluated to be

$$\kappa_{em}^2 = \frac{(Bl)^2}{k_{sp} L}, \tag{4.21}$$

where B is the magnetic flux density, k_{sp} the spring inductance, l is the length of the wire in the coil, and L the coil inductance. In the case of piezoelectric generators, the coupling factor was evaluated as

$$\kappa_{em}^2 = \frac{dE}{\epsilon}, \tag{4.22}$$

where d is the piezoelectric strain coefficient, E is Young's modulus, and ϵ the dielectric permittivity. Electrostatic transducers are nonlinear, and consequently the coupling coefficient depends on the amplitude of the input excitation. Based on the preceding analysis, it was determined in [100] that comparable efficiencies and maximum generated power can be obtained for electromagnetic and piezoelectric devices. In the case of electrostatic devices, the comparison is more difficult due to the nonlinear nature of the devices, but it was also determined that for a given input excitation amplitude, comparable coupling coefficients may also be obtained.

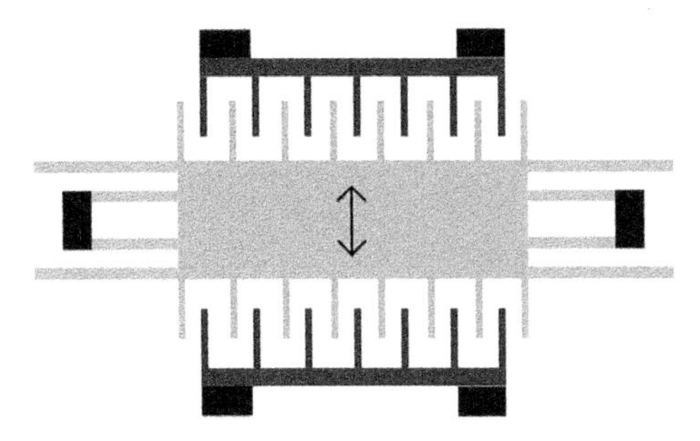

Figure 4.6 MEMS-based electrostatic energy harvester layout representation, based on [90].

4.6 Vibration Energy Harvester Examples

There are many kinetic energy harvesting device examples in the literature, as the fundamental physics behind the kinetic-to-electrical energy conversion has been studied for many years. As an example, it has been already noted in Section 4.2.2 that Tesla's electromagnetic dynamo transducer was patented in 1887 [93]. Technology, however, is making possible new implementations especially tailored toward microenergy scavenging or harvesting applications.

Silicon micromachining MEMS technology is particularly suitable for implementing mechanically tunable capacitors for electrostatic transducers due to its compatibility for integration with silicon microelectronics and its availability for mass production [91]. The first MEMS-based electrostatic microenergy generator in the literature was reported in [90]. The simulated harvested energy by the transducer was 8.6 μW, whereas approximately 5.6 μW would be usable and the rest would be dissipated by the controller. A view of the capacitor layout is shown in Figure 4.6. The capacitor comprises two anchored stationary comb terminals and an oscillating mass in the middle. This topology is known as in-plane overlap type, where the capacitance changes by varying the overlapping between the fingers. There exist two additional topologies, the in-plane gap closing, a conceptual view of which is shown in Figure 4.7, and the out-of-plane gap closing shown in Figure 4.8 [91, 99].

A characteristic example of a piezoelectric-based vibration energy harvester is the one developed by Roundy and Wright [101]. The structure comprises a piezoelectric bimorph with an attached mass M, shown in Figure 4.9. A bimorph is a cantilever that comprises two sheets of piezoelectric material sandwiched together with a center nonpiezoelectric shim and bonded together. As the cantilever bends, one of the piezoelectric sheets is stretched and the other is compressed according to the direction of the bending force. The bimorph is operated in the 31 mode, exploring the strain d_{31} tensor element where the stress is applied in the z direction and the electric field is generated along the x direction, as explained in Section 4.2.3. A prototype harvester was fabricated

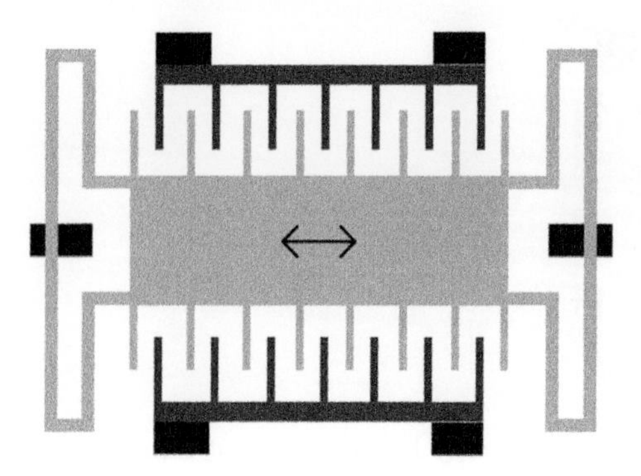

Figure 4.7 In-plane gap closing layout representation, based on [91, 99].

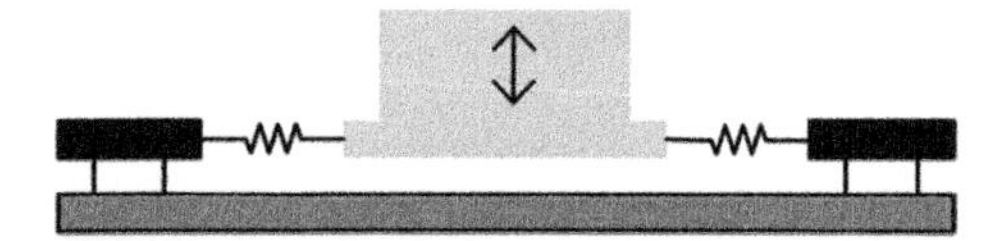

Figure 4.8 Out-of-plane gap closing layout representation, based on [91, 99].

where the bimorph was made with PZT-5A sheets sandwiched together with a steel shim and the mass was made from an alloy of tin and bismuth [101]. The structure was designed to resonate at 120 Hz and was tested with external vibrations with acceleration of 2.5 ms^{-2}. The prototype delivered about 80 μW at a 250 KΩ load. Subsequent further optimized designs were able to deliver up to 375 μW under the same excitation of 2.5 ms^{-2} at 120 Hz.

Shenk and Paradiso [10] demonstrated a body-worn piezoelectric energy harvester placed in the insole of a shoe. The concept is shown in Figure 4.10, which illustrates two possible topologies for the harvester, a semiflexible PZT-based bimorph placed under the heel or a flexible PVDF bimorph placed under the ball of the foot. The PVDF stave was able to harvest 1.3 mW on average in a 250 KΩ load at a 0.9 Hz walking pace. The PZT dimorph produced on average 8.4 mW in a 500 KΩ load under approximately the same walking pace. The harvester was used to power an active RFID tag operating at 310 MHz. Subsequently, Orecchini et al. demonstrated an RFID tag powered by a shoe-mounted piezoelectric energy

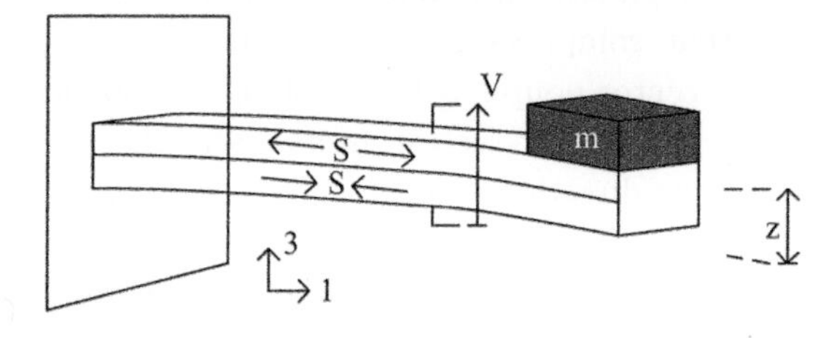

Figure 4.9 Piezoelectric bimorph energy harvester, based on [101].

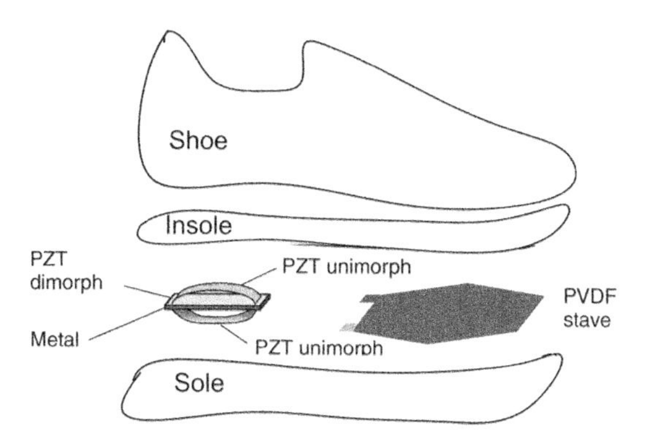

Figure 4.10 Schematic representation of a piezoelectric bimorph shoe-mounted energy harvester based on [10].

Figure 4.11 Schematic representation of shoe-mounted inkjet-printed RFID tag based on [102].

harvester where the electronic circuitry and the antenna were inkjet printed on a flexible paper substrate (see Figure 4.11) [102].

The world's first watch powered by an electromagnetic type vibration energy harvester was introduced in 1988 by Seiko [103]. The movement of the arm on which the watch is worn caused a rotor to rotate and subsequently spin an electromagnetic vibration–based harvester. The harvester is capable of powering the watch, which required only 0.7 μW.

An example of a miniature silicon-based electromagnetic-type vibration harvester is the one developed by Beeby et al. [104]. The device comprised four magnets and a coil implemented in a vibrating silicon cantilever paddle layer. The harvester was able to generate 21 nW from an external source with acceleration of 1.92 ms$^-$2 at 9.5 KHz [99].

Finally, a hybrid system combining an electromagnetic-type vibration energy harvester with a radio frequency (RF) energy harvester was introduced by Gu et al. [105]. A schematic of the circuit is shown in Figure 4.12. The output of an antenna is connected to a shunt diode rectifier circuit harvesting RF power and converting it to dc electrical power. The output of an electromagnetic-type vibration harvester comprising a coil placed inside the magnetic field of a permanent magnet is electrically connected in parallel to the shunt diode, which also rectifies it to produce a combined dc output current delivered to a load. The cooperating functionality of the two harvesters results in improved efficiency for the combined system. The implementation of hybrid harvesting systems provides an added challenge to the designer as the electrical connection

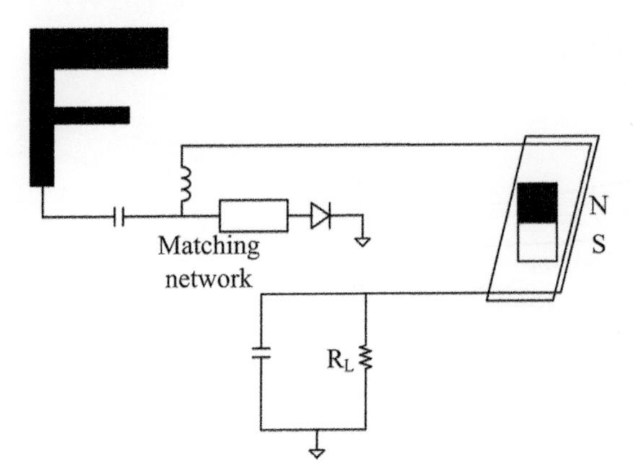

Figure 4.12 Schematic representation of hybrid vibration and an RF energy harvester based on [105].

of the two individual harvesters results in loading each other and consequently affecting their operating efficiency.

4.7 Problems and Questions

1. How many types of electrostatic transducers exist and what is their principle of operation?
2. What is the operating principle of electromagnetic transducers?
3. What is the operating principle of piezoelectric transducers? Name a ceramic and a polymer piezoelectric material. What does a piezoelectric strain constant d_{31} refer to?
4. Describe the model of vibration energy harvesting transducers proposed by Williams and Yates.
5. Order the various vibration sources listed in Table 4.1 according to the maximum available energy for harvesting assuming that the same harvesting transducer is installed on all of the sources.
6. If the available energy for harvesting by a constant voltage electrostatic energy harvester is 5 μW, how much is the required capacitance variation when the harvester capacitor is charged to 10 V?
7. What is the available energy for harvesting by a constant charge electrostatic energy harvester if the same capacitance variation is used as in the previous problem and the capacitor is first charged to a voltage of 5 V at a minimum capacitance and the maximum voltage across its plates is 10 V when the maximum capacitance is reached?
8. An electromagnetic transducer comprises a square coil with N = 20 turns and has a side with length 10 cm. The coil is allowed to vibrate along its axis x inside a magnetic field B that is also directed along axis x. The magnetic

field has a slope of $dB/dx = 0.1$ T at the location of the coil. The vibration source makes the coil vibrate with an acceleration magnitude of 10 m/s^2 at a frequency of 2 Hz. What is the maximum open-circuit voltage magnitude at the coil terminals?

5 Thermal Energy Harvesting

5.1 Introduction

Thermoelectric transducers convert thermal energy into electric energy. Thermal energy is generated as a result of a multitude of phenomena and applications, in some cases intentionally but most of the time as waste heat from a process or reaction, from industrial plants to buildings, heating systems, and automobiles to the human body, which, in turn, provide numerous applications for thermal energy harvesters.

This chapter begins with a description of thermoelectric phenomena and the geometrical structure of a thermoelectric generator (TEG). Next an introduction to the theory of heat transfer is presented in order to provide the theoretical background for the analysis of the performance of thermoelectric generators, and it is followed by theoretical expressions for the efficiency of TEGs. The next section deals with the figure of merit of different thermoelectric materials, and it is followed by a SPICE model of a thermoelectric generator. Finally the chapter ends with selected examples of TEG systems.

5.2 Thermoelectric Phenomena

There are three thermoelectric phenomena that govern the conversion of thermal energy to electrical energy and vice versa: (a) the Seebeck effect, (b) the Peltier effect, and (c) the Thomson effect. In the following, a summary of the three effects is provided.

5.2.1 The Seebeck Effect

According to the Seebeck effect, a temperature gradient between two different metals or semiconductors that are in contact creates a voltage difference between the two components [63]. Given a set of two different metal or semiconductor materials 1 and 2 that are connected forming two junctions as shown in Figure 5.1a, the presence of different temperatures T_H and T_C at the two junctions results in a voltage V_{oc} across the two contacts.

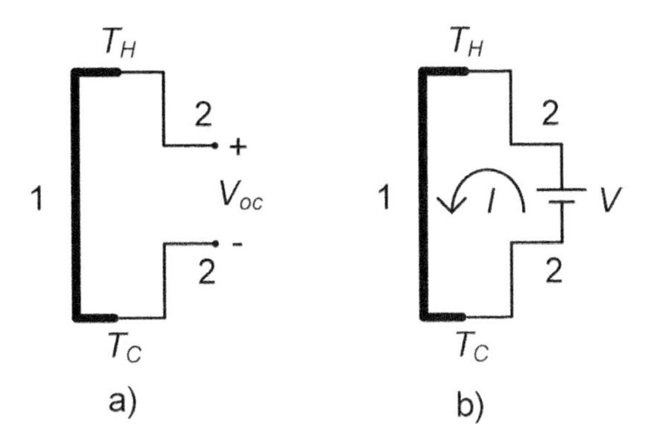

Figure 5.1 Thermoelectric phenomena: (a) the Seebeck effect and (b) the Peltier effect.

The voltage is given by

$$V_{oc} = \alpha_{12}\Delta T, \tag{5.1}$$

where $\Delta T = T_H - T_C$ and the coefficient α_{12} is the called the Seebeck coefficient and has units of V/K. The Seebeck coefficient depends on both materials 1 and 2 and can be negative or positive. As we will see in a later section, the Seebeck coefficient is higher when semiconductor materials rather than metals are used to form the junctions, and consequently TEG devices typically are made of semiconducting materials.

5.2.2 The Peltier Effect

The Peltier effect is the inverse of the Seebeck effect [63]. In other words, the application of an external voltage difference V at the junctions of two different metals or semiconductors results in a current I flowing through the junctions, which, in turn, results in one junction absorbing thermal energy and the other junction generating thermal energy. As a result of the current flow, a heat flow rate Q is created. Consequently, a temperature gradient is generated between the junctions (Figure 5.1b).

The Peltier effect is described by the following equation

$$Q = \pi_{12}I, \tag{5.2}$$

where the heat flow rate between the two junctions is measured in W and the current through the circuit in A. The coefficient π_{12} is called the Peltier coefficient and has units of W/A or equivalently V. The Peltier coefficient, like the Seebeck coefficient, is a relative coefficient corresponding to the two materials forming the junctions.

5.2.3 The Thomson Effect

The third thermoelectric effect is the Thomson effect, which occurs in one material (metal or semiconductor) when its edges are subject to a temperature difference and at the same time they are subject to a voltage difference resulting in current flowing through the material [63]. As a result, there is heat Q_T absorbed or dissipated by the material, which depends both on the applied current and the temperature difference. The Thomson effect is described by the equation

$$Q_T = \beta I \Delta T, \tag{5.3}$$

where β is the Thomson coefficient, which is measured in $WI^{-1}K^{-1}$ or equivalently VK^{-1}. The heat rate due to the Thomson effect is smaller than the one due to the Peltier effect; nonetheless, it can become significant when the temperature difference ΔT is large [63].

5.2.4 The Kelvin Relationships

The three thermoelectric coefficients a_{12}, π_{12}, and β are related by the Kelvin relationships [63]

$$\pi_{12} = \alpha_{12} T \tag{5.4}$$

and

$$\frac{d\alpha_{12}}{dT} = \frac{\beta_1 - \beta_2}{T}. \tag{5.5}$$

The first equation relates the Seebeck and Peltier effects and demonstrates the reversible nature of the effects and the fact that the same set of materials is suitable both for electric power generation and for thermal power generation (or refrigeration). The second equation relates the Seebeck effect with the Thompson effect and enables the definition of an absolute Seebeck coefficient for a single material as

$$\alpha = \int \frac{\beta}{T} dT. \tag{5.6}$$

The Seebeck coefficient α_{12} corresponding to the two junctions of the two materials 1 and 2 is proven to be equal to the difference between the absolute Seebeck coefficients of each of the two materials $\alpha_{12} = \alpha_1 - \alpha_2$ [63]. Similarly, once an absolute Seebeck coefficient is defined, (5.4) defines an absolute Peltier coefficient as $\pi_{12} = \pi_1 - \pi_2$. When the magnitude of the Seebeck or Peltier coefficients of the two materials is equal, $\alpha_1 = -\alpha_2$ and $\pi_1 = -\pi_2 = \pi$, then $\alpha_{12} = 2\alpha$ and $\pi_{12} = 2\pi$.

5.3 Thermoelectric Generators

TEGs are typically constructed by forming matrices of pairs of p-type and n-type semiconductor columns called pellets. The pellets are electrically connected in

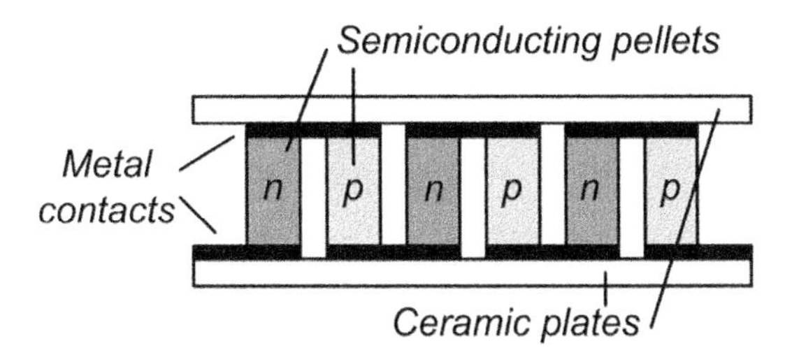

Figure 5.2 Cross-section of a TEG.

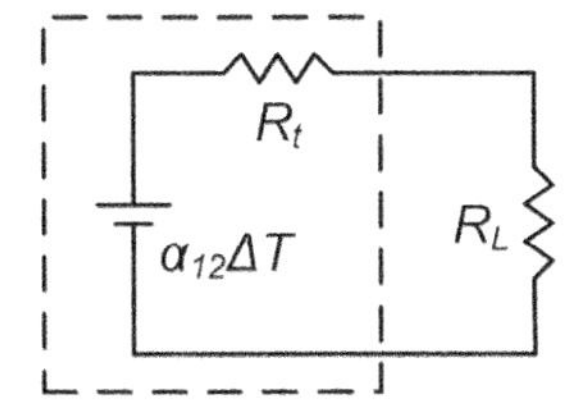

Figure 5.3 Electrical equivalent circuit of a TEG.

series using conducting (for example, copper or aluminum) strips, and are sandwiched between thermally conductive ceramic plates, as shown in Figure 5.2.

The output voltage of TEGs depends on the size and number of pellet pairs and typically ranges from 10 to 50 mV/K [106]. Due to the fact that the output voltage of a TEG in a energy harvesting application scenario takes such small values, a TEG is usually connected to a load consisting of a switched-type voltage converter, such as a boost or fly-back converter in order to produce a desired voltage required by standard circuitry [106].

The electrical equivalent circuit of a TEG consists of a Seebeck voltage source with a value V_{oc} given by (5.1) in series with an electrical resistance R_t representing the heat generated inside the TEG due to thermal losses as the electrical current flows through the pellets. The equivalent circuit is shown in Figure 5.3.

5.4 Heat Transfer Fundamentals

Heat is transferred through three physical mechanisms, *conduction*, *convection*, and *radiation* [107]. Each mechanism is governed by a rate equation, which provides a quantification of the heat flux rate measured in $\mathrm{W/m^2}$.

Conduction is the transfer of energy between hot and cold particles of a material. The rate equation of conductive heat transfer is Fourier's law. Conductive heat transfer typically occurs in TEGs, and for this reason Fourier's law is described in more detail in the next subsection.

Convection is the transfer of heat due to the bulk, macroscopic motion of fluids. In the case of solids, an important scenario is that of a fluid with temperature T_∞

flowing over a heated solid surface of temperature T_s. In this case, the convection rate equation is

$$q_{cv} = h\,(T_s - T_\infty)\,. \tag{5.7}$$

The parameter h (W m^{-2}K^{-1}) is the convection heat transfer coefficient.

Finally, all matter at a temperature higher than zero Kelvin emits thermal radiation whose upper limit is given by the rate equation known as the Stefan–Bolzmann law

$$q_r = \sigma T^4, \tag{5.8}$$

where q_r is the radiated heat flux rate of an ideal blackbody radiator measured in W/m^2 that is at temperature T. The constant of proportionality σ is the Stefan–Bolzmann constant, which is equal to $\sigma = 5.67 \cdot 10^{-8}$ W m^{-2}K^{-4}.

5.4.1 Fourier's Law

The conduction heat transfer process is described by Fourier's law expressed by

$$q = -k\nabla T. \tag{5.9}$$

According to Fourier's law, the heat flux rate q measured in W/m^2 is proportional to the temperature gradient ∇T (K m^{-1}). The constant of proportionality k is the thermal conductivity of the material (W m^{-1} K^{-1}). The minus sign expresses the fact that heat is transferred from points of the material with a higher temperature toward points of lower temperature. A TEG can be approximately modeled as a one-dimensional (1D) problem of heat diffusion from the hot plate toward the cold plate (Figure 5.2). In this case, Fourier's law becomes

$$q = -k\frac{dT}{dx}, \tag{5.10}$$

where x represents the vertical direction between the two plates shown in Figure 5.2. Once the steady-state temperature distribution in the material is defined, the heat flux can be computed using Fourier's law.

In the case of 1D problems, Fourier's law presents an analogy with Ohm's law of electrical circuits [107]. Let us consider an infinitesimal volume $V = AL$, where the heat rate is flowing through the volume surface A and the temperature gradient is taken along the length L. Then we can write Fourier's law as

$$Q = -\frac{kA}{L}(-dT) \Rightarrow Q = K\Delta T \Rightarrow Q\Theta = \Delta T, \tag{5.11}$$

where $K = kA/L$ is defined as the thermal conductance (W K^{-1}) and $\Theta = 1/K$ is the thermal resistance. The analogy with Ohm's law $IR = \Delta V$ is now obvious. This fact is explored in computing the steady state of TEGs using electrical circuit simulators, as we will see in Section 5.6.

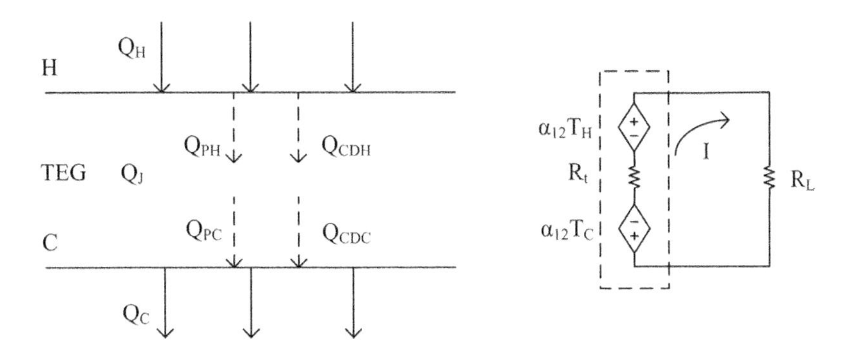

Figure 5.4 Conservation of energy in a TEG.

5.4.2 The First Law of Thermodynamics

In heat transfer problems, one must always apply the first law of thermodynamics, in other words the law of conservation of energy [107]. One may express this law in different formats, for example in terms of energy flux rates or in terms of energy rates (i.e., power). We assume that the heat distribution is uniform over the boundary surfaces of the TEG, and in this case we can convert heat flux rates q to heat rates Q simply by multiplying the former by the area A of the surface of the TEG boundary, i.e., $Q = qA$. The conservation of energy for the TEG system in terms of heat rates is depicted in Figure 5.4, and it takes the form

$$Q_{st} = Q_H - Q_C + Q_J. \tag{5.12}$$

Equation (5.12) states that the stored thermal and mechanical energy rate Q_{st} in the TEG system of a certain volume is equal to the difference between the inflow of energy rate Q_H and outflow of energy rate Q_C at the system boundary surface, plus any thermal energy rate Q_J generated in the system.

When we are dealing with the boundary surfaces of the system, conservation of energy across a boundary is expressed by the fact that the incoming energy rate must be equal to the outgoing energy rate at each boundary. Based on Figure 5.4, the conservation of energy in the hot H boundary becomes

$$Q_H = Q_{PH} + Q_{CDH}, \tag{5.13}$$

where Q_{PH} is the absorbed Peltier heat and Q_{CD} contains contributions from heat conduction and Joule heat generation in the TEG, which we derive using the heat diffusion equation in the next section. The conservation of energy in the cold C junction becomes

$$Q_C = Q_{PC} + Q_{CDC}, \tag{5.14}$$

where Q_{PH} is the emitted Peltier heat and Q_{CD} contains contributions from heat conduction and Joule heat generation in the TEG accordingly. The magnitude of Q_{PH} is different from Q_{PC} because of the different temperatures at the H and C boundaries. Furthermore, Q_{CDH} is different from Q_{CDC} because the heat conduction rate flows from the H to the C boundary, but the generated Joule heat is emitted from both boundaries equally as we will see in the next sections.

5.4.3 The Heat Diffusion Equation

Once the steady-state temperature distribution in a material is determined, one can compute the heat flow using the rate equations. The heat diffusion equation, or simply the heat equation, provides a means to determine the temperature distribution as a function of time. In Cartesian coordinates, it is given

$$\nabla(k\nabla T) + q_j = \rho c_p \frac{\partial T}{\partial t}, \tag{5.15}$$

where k is the thermal conductivity of the material, ρ is the density (kg m^{-3}), and c_p (J kg^{-1} K^{-1}) is the specific heat of the material. The product ρc_p is called volumetric heat capacity and measures the ability of the material to store thermal energy [107]. Finally, q_j is the heat density rate measured in Wm^{-3}. In the case of a one-dimensional conduction problem and of a material with a constant thermal conductivity, which represents an approximate model for the TEG, the heat equation becomes

$$\frac{\partial^2 T}{\partial x^2} + \frac{q_j}{k} = \frac{1}{\alpha}\frac{\partial T}{\partial t}, \tag{5.16}$$

where $\alpha = k/(\rho c_p)$ is the thermal diffusivity. The steady-state condition is obtained by setting the partial derivative of the temperature versus time equal to zero, in which case one obtains

$$\frac{\partial^2 T}{\partial x^2} + \frac{q_j}{k} = 0. \tag{5.17}$$

5.5 TEG Efficiency

The conversion efficiency η of a TEG is defined as the ratio of the electrical power delivered to a load R_L connected to the TEG P_L divided by the heat rate absorbed at the hot junction Q_H

$$\eta = \frac{P_L}{Q_H}, \tag{5.18}$$

where Q_H is the heat rate at the hot junction. Assuming a uniform heat distribution along the junction of the TEG, the heat rate is $Q_H = q_h A$, where q_h is the heat flux rate and A is the surface area of the hot junction.

5.5.1 The Carnot Efficiency

Let's consider first the ideal case where only the Peltier and Seebeck phenomena exist. We do not consider heat conduction phenomena and no heat generation inside the TEG. The first assumption results in the thermal conductivity of the TEG being zero $k = 0$, while the second assumption results in the electrical resistance of the TEG being set to zero $R_t = 0$ in the model of Figure 5.3.

The power delivered to the load is $P_L = I^2 R_L$, where I is the current flowing through the TEG and the load R_L. Using (5.1), one has

$$V_L = \alpha_{12}(T_H - T_C) \tag{5.19}$$

$$I_L = \frac{\alpha_{12}(T_H - T_C)}{R_L} \tag{5.20}$$

and

$$P_L = I_L^2 R_L = \alpha_{12}^2 (T_H - T_C)^2 \frac{1}{R_L}. \tag{5.21}$$

Application of the first law of thermodynamics at the hot junction boundary of the TEG results in the heat rate Q_H absorbed at the hot junction being equal to the Peltier absorbed heat rate, and therefore

$$Q_H = \pi_{12} I_L = \alpha_{12} T_H I_L, \tag{5.22}$$

where (5.2) and (5.4) were used. The TEG efficiency becomes

$$\eta_C = \frac{T_H - T_C}{T_H}. \tag{5.23}$$

This is known as the Carnot efficiency and it represents an upper bound in the efficiency of TEGs. One should highlight that even the ideal Carnot efficiency takes very small values in many application scenarios of energy harvesting, for example a hot junction with tempetature $\Delta T = 10$ K above a cold junction at room temperature $T_C = 300$ K gives a Carnot efficiency of 3.33%.

5.5.2 Conversion Efficiency Considering Heat Conduction and Thermal Losses in the TEG

An ideal TEG presents no thermal conductivity and no electrical resistance. All materials, however, present a nonzero thermal conductivity k and a nonzero electrical resistance R_t.

The power delivered to the load in the case of a nonzero electrical resistance R_t as shown in Figure 5.3 becomes

$$V_L = \alpha_{12}(T_H - T_C)\frac{R_L}{R_t + R_L} \tag{5.24}$$

$$I_L = \frac{\alpha_{12}(T_H - T_C)}{R_t + R_L} \tag{5.25}$$

and

$$P_L = I_L^2 R_L = \alpha_{12}^2 (T_H - T_C)^2 \frac{R_L}{(R_t + R_L)^2}. \tag{5.26}$$

The load power becomes maximum when $R_t = R_L$, in which case

$$P_{Lmax} = \frac{\alpha_{12}^2 (T_H - T_C)^2}{4R_t}. \tag{5.27}$$

In order to compute the heat rate at the hot junction, we apply the heat diffusion equation to determine the temperature distribution along the TEG height. The heat diffusion equation in the steady state $\partial T/\partial t = 0$ takes the form

$$\frac{\partial^2 T}{\partial x^2} + \frac{q_j}{k} = 0. \tag{5.28}$$

Considering a uniform heat generation along the TEG, the Joule heat density q_j takes is equal to

$$q_i = \frac{P_J}{V} = \frac{I^2 R_t}{AL}, \tag{5.29}$$

where A is the area of one pellet pair, L its length, and $V = AL$ its volume.

The solution of (5.28) has the form

$$T(x) = -\frac{q_i}{2k} x^2 + C_1 x + C_2. \tag{5.30}$$

The temperature variation across the TEG pellet height has a parabolic profile. The constants C_1 and C_2 are determined by the boundary conditions

$$\begin{aligned} T(0) &= T_H \\ T(L) &= T_C, \end{aligned} \tag{5.31}$$

where L is the length (or height) of the pellet. One can easily compute

$$T(x) = -\frac{I^2 R_t}{2ALk} x^2 + \left(-\frac{T_H - T_C}{L} + \frac{I^2 R_t}{2Ak} \right) x + T_H. \tag{5.32}$$

Once we have determined the temperature distribution $T(x)$, application of Fourier's law (5.11) at the hot junction of the TEG gives the conduction heat flux q_{cdh}.

In order to find the heat rate at the hot junction H ($x = 0$) of the TEG as shown in Figure 5.4, first we determine the first derivative of the temperature from (5.32):

$$\frac{dT(x)}{dx} = -\frac{I^2 R_t}{ALk} x - \frac{T_H - T_C}{L} + \frac{I^2 R_t}{2Ak}. \tag{5.33}$$

Then Fourier's law (5.11) gives

$$q_{cdh} = -k \left(\frac{dT}{dx} \right)_{x=0} = k \frac{T_H - T_C}{L} - \frac{I^2 R}{2A}. \tag{5.34}$$

The conduction heat rate Q_{CDH} is be computed from the heat flux q_{cdh} using the cross-section A of the pellet pair as $Q_{CDH} = q_{cdh} A$, resulting in

$$Q_{CDH} = \frac{kA}{L}(T_H - T_C) - \frac{I^2 R}{2} = K(T_H - T_C) - \frac{I^2 R}{2}, \tag{5.35}$$

where K is the thermal conductivity of a pellet pair with height L and cross-section A. Q_{CDH} included both the conduction heat rate, leaving the hot junction and a Joule heat term generated in the TEG.

We have seen in (5.13) that the application of the first law of thermodynamics (5.12) at the hot junction boundary gives that the incoming absorbed heat rate Q_H is equal to the Peltier emitted heat rate $Q_{PH} = \alpha_{12} T_H I$ plus the heat conduction and Joule heat generation term Q_{CDH}:

$$Q_H = Q_{PH} + Q_{CDH} = \alpha_{12} T_H I + K(T_H - T_C) - \frac{I^2 R}{2}. \tag{5.36}$$

Following a similar calculation, we can show that the emitted heat rate Q_C at the cold junction boundary of the TEG is

$$Q_C = \alpha_{12} T_C I + K(T_H - T_C) + \frac{I^2 R}{2}. \tag{5.37}$$

The analysis has been made for one pellet pair. A TEG typically has a large number N of pellet pairs that are connected thermally in parallel and electrically in series (Figure 5.2). In this case, the thermal conductance K, the electrical resistance R_t, the Seebeck voltage V_{oc}, and consequently the absorbed Q_H and emitted heat rates Q_C are all multiplied by the number of pellet pairs N.

We have seen in (5.18) that in order to calculate the efficiency of the TEG we only require the absorbed heat rate Q_H at the hot junction. Using (5.18), (5.25) and (5.36) the conversion efficiency of the TEG becomes

$$\eta = \frac{\alpha_{12}^2 (T_H - T_C) R_L}{\alpha_{12}^2 \left[T_H (R_t + R_L) - \frac{\Delta T R_t}{2}\right] + K(R_t + R_L)^2} \tag{5.38}$$

or

$$\eta = \eta_C \frac{Z R_L}{Z \left[(R_t + R_L) - \frac{n_C R_t}{2}\right] + \frac{1}{R_t T_H}(R_t + R_L)^2}, \tag{5.39}$$

where η_C is the Carnot efficiency and $Z = \alpha_{12}/(K R_t)$ is defined as the figure of merit of the thermoelectric material of the TEG. The figure of merit is proportional to the square of the Seebeck coefficient, and inversely proportional to the thermal conductivity and the electrical resistance of the TEG material and has units K^{-1}. Good thermoelectric materials have a large Seebeck coefficient but also low thermal conductivity and low electrical resistance. It is customary to multiply Z with the temperature in order to obtain a unitless parameter.

As we have seen, the condition for maximum delivered electrical power to the load is $R_t = R_L$. However, due to the dependance of both the load power P_L and the absorbed power Q_H on the load resistance R_L, the load value that provides maximum efficiency η is different from the load value that provides maximum delivered power. Taking $\partial \eta / \partial R_L = 0$, one obtains the load value that corresponds to maximum efficiency as

$$R_{Lm} = R_t \sqrt{1 + Z\bar{T}}, \tag{5.40}$$

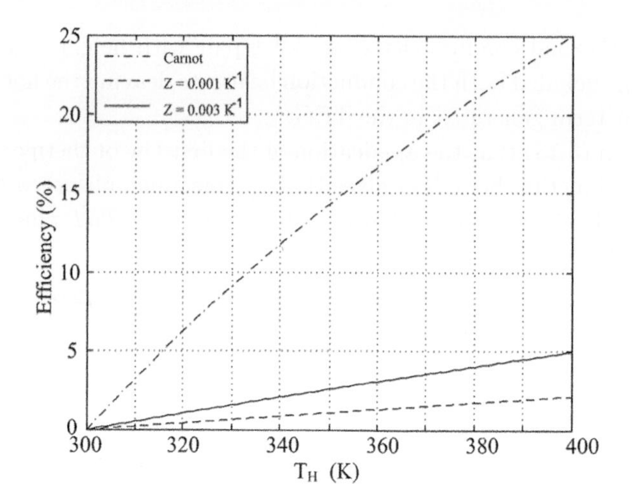

Figure 5.5 TEG conversion efficiency ($T_C = 300$ K).

where $\bar{T} = (T_H + T_C)/2$. The maximum efficiency then becomes

$$\eta = \eta_C \frac{\sqrt{1 + Z\bar{T}} - 1}{\sqrt{1 + Z\bar{T}} + \frac{T_C}{T_H}}. \tag{5.41}$$

The conversion efficiency for different values of the figure of merit versus the hot side temperature is shown in Figure 5.5, when the cold side is at $T_C = 300$ K. The Carnot efficiency is also included for comparison. It should be noted that for a TEG made of a material with figure of merit $Z = 0.003$ K^{-1}, when the hot side is at 10 K above room temperature, the conversion efficiency is just 0.52%.

5.5.3 The Figure of Merit

The figure of merit is proportional to the square of the Seebeck coefficient, inversely proportional to the thermal conductivity and the electrical resistance of the TEG material, and has units K^{-1}. Good thermoelectric materials have a large Seebeck coefficient but also low thermal conductivity and low electrical resistance.

All three parameters depend on the carrier concentration in the material. The dependence is pictured in Figure 5.6 [108]. The results show that highly doped semiconductor materials are more suitable for thermoelectric applications than metals or insulators.

Furthermore, Figure 5.7 [63] shows obtained values of the figure of merit for several materials. The figure of merit Z is normalized to the temperature in order to obtain a unitless parameter. One can see that Bi_2Te_3 has a maximum figure of merit value around room temperature $T = 300$ K, and for this reason it is commonly used in TEG applications. Typical values for the figure of merit range around $2.5 \cdot 10^{-3}$ K^{-1} to $3 \cdot 10^{-3}$ K^{-1}.

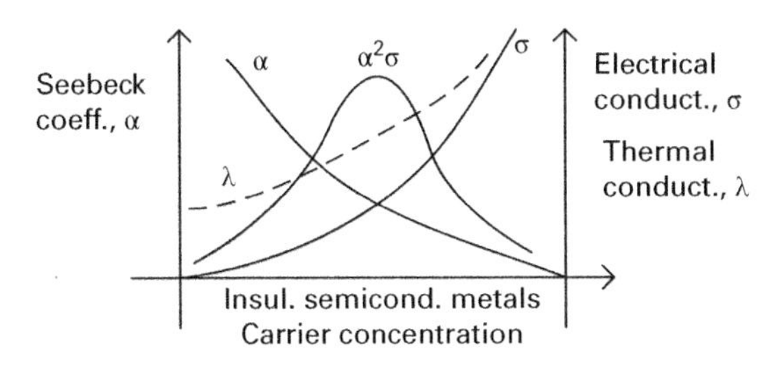

Figure 5.6 Schematic description of Seebeck coefficient, thermal conductivity, and electrical resistance dependence on carrier concentration based on [108].

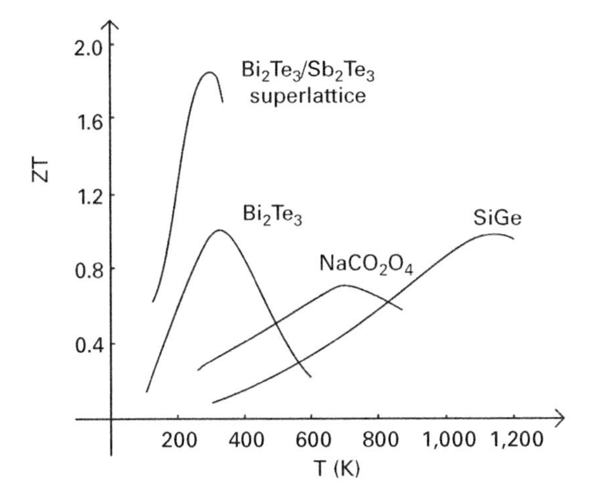

Figure 5.7 Selected normalized figure of merit ZT plots for different thermoelectric materials, reproduced from [63].

5.6 A Thermal and Electrical SPICE Model for the TEG

Fourier's law for heat conduction presents an analogy with Ohm's law in electrical circuits. Due to this fact, it is possible to model and analyze thermal problems using electrical simulators such as Simulation Program with Integrated Circuit Emphasis (SPICE). The analogy between the various thermal and electrical quantities is summarized in Table 5.1 [109].

Table 5.1 Analogy between thermal and electrical quantities [109].

Thermal	Unit	Electrical	Unit
Heat rate, Q	W	Current, I	A
Temperature, T	K	Voltage, V	V
Thermal resistance, $\Theta = 1/K$	K W^{-1}	Resistance, R	Ω
Heat capacity, C	J K^{-1}	Capacitance, C	F
Absolute zero temperature	0 K	Ground	0 V

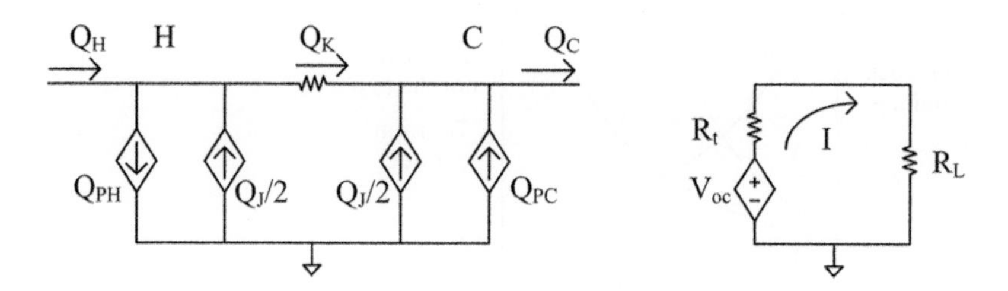

Figure 5.8 SPICE model of a TEG [109].

Using Table 5.1, the energy balance equations at the hot and cold boundaries of the TEG (5.36) and (5.37) can be represented with controlled current sources and a resistor. Thus, it is possible to build a SPICE equivalent model for the TEG [109] shown in Figure 5.8, where

$$\begin{aligned} Q_{PH} &= \alpha_{12} T_H I \\ Q_{PC} &= \alpha_{12} T_C I \\ Q_J &= I^2 R_t \\ Q_K &= K(T_H - T_C) \\ V_{oc} &= \alpha_{12}(T_H - T_C). \end{aligned} \tag{5.42}$$

The model can be used to find the steady state of the TEG. The model parameters can be computed from the TEG manufacturer specifications [109]. It is further possible to introduce a thermal capacity C (J/K) in the SPICE model, which allows to study the transient behavior of the TEG [110].

5.7 Thermal Energy Harvester Systems

Seiko presented in 1998 the first watch that was powered by a thermoelectric transducer [111]. The power that is required in order to operate a quartz digital wristwatch is approximately 20–40 μW [63]. There are many application scenarios where electronic circuits generate a large amount of heat. For example, typically in RF and microwave electronics a power amplifier operates with low efficiency in order to maintain an acceptable level of distortion and consequently a significant fraction of the power used to supply the power amplifier is dissipated in heat. Therefore, it is possible to use a TEG in order to convert some of the wasted heat back into electrical power. Such an application scenario has been studied in [112]. A 1.37 W power amplifier with 10 dB of gain was used, operating at 2.45 GHz with a measured power-added efficiency (PAE) of 34%. The PAE is defined as $(P_o^{RF} - P_i^{RF})/P_{dc}$, which means that the amplifier generated approximately 0.88 W of heat.

The amplifier printed circuit board with a commercial TEG placed below is shown in Figure 5.9. The electronics of the TEG board are shown in the figure

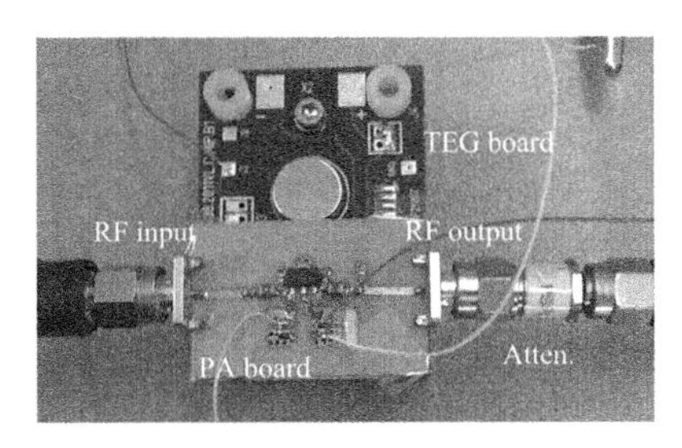

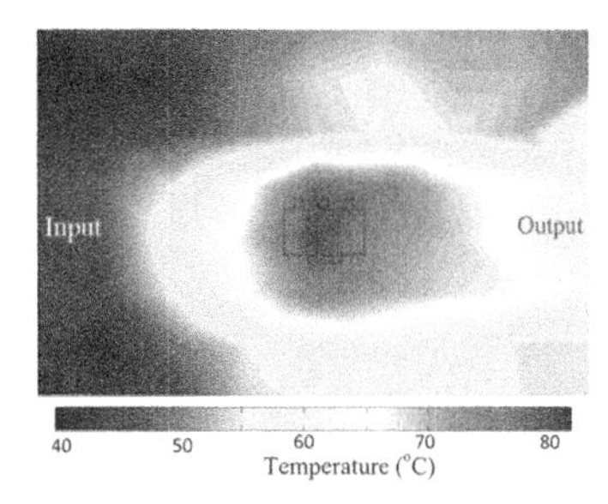

Figure 5.9 Thermal energy harvesting from an RF power amplifier. ©2013 IEEE. Reprinted with permission from [112]

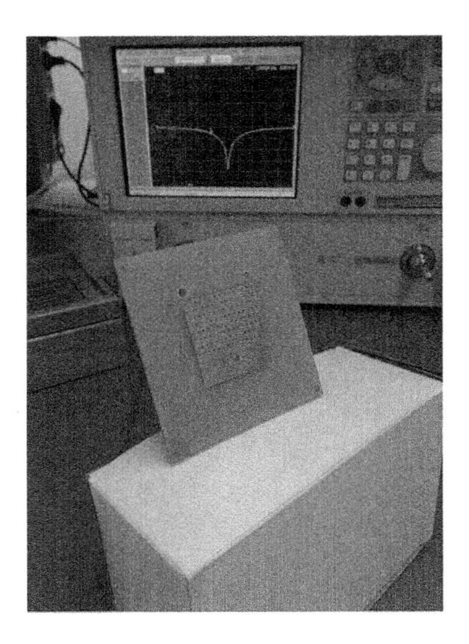

Figure 5.10 Patch antenna integrating a thermoelectric generator [113]. Photo courtesy of Dr. Marco Virili, Qorvo Inc.

whereas the hot side of the TEG was placed in contact with the ground place of the amplifier board directly below the amplifier packaged integrated circuit. A measured temperature map of the power amplifier printed circuit board when the amplifier was operating is also shown in Figure 5.9. At steady state, the temperature at the hot side of the TEG was measured to be 313.1 K, whereas the temperature at the cold side was 305.9 K. The temperature difference of 7.2 K was maintained with the help of a heat sink placed below the TEG. This temperature gradient corresponds to a Carnot efficiency of 2.3%. A measured output power of 1 mW was obtained from the TEG, which, although represents a very low efficiency for the thermoelectric generator, as an absolute value it is sufficient to power a wireless sensor circuit performing some monitoring function, for example.

The integration of multiple harvesting systems of different technologies is important in order to optimize the energy autonomy of a wireless sensor circuit by

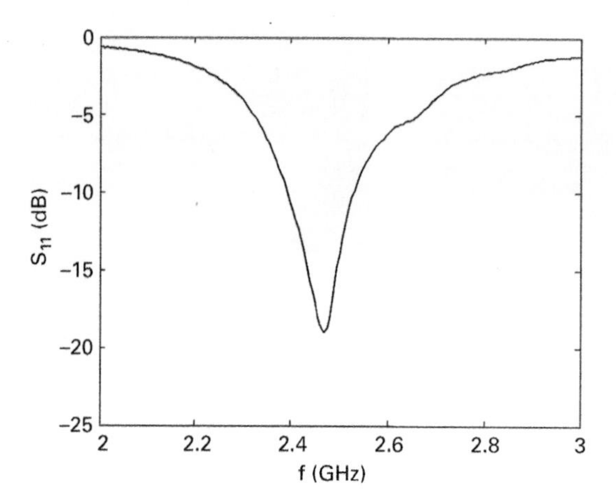

Figure 5.11 S-parameters of the patch antenna with TEG [113].

exploring different sources of power. Due to the typically small size of the sensors, the integration becomes a challenge in order to minimize the used space. In [113], it was investigated whether a TEG can be integrated with an antenna, which can be a communication antenna or an RF energy harvesting antenna. A commercial TEG was placed above a quarter-wave shorted patch antenna implemented in FR4 substrate. The patch antenna with and without the TEG placed on top is shown in Figure 5.10. A shorted patch antenna design was selected in order to provide a low thermal resistance connection between one of the TEG surfaces and the ground plane of the antenna.

The antenna dimensions were retuned with the help of commercial electromagnetic simulator software in order for the antenna to operate in the 2.4 GHz industrial, scientific, and medical (ISM) band. The measured s-parameters of the antenna prototype are shown in Figure 5.11, where we can see that the desired operating bandwidth is obtained. The measured radiation pattern of the antenna with the TEG at 2.45 GHz is shown in Figure 5.12, which showed an obtained gain of approximately 2.3 dB. The presence of the TEG reduced the antenna gain by less than 1 dB.

Following the successful implementation of the patch antenna with the TEG, a shorted patch antenna integrating both a TEG and a solar cell on top was successfully demonstrated in [114]. The antenna prototype is shown in Figure 5.13, whereas its measured performance is shown in Figure 5.14 verifying that with proper design the presence of the TEG and the solar cell has a minimal effect in the operation of the parch antenna. Such systems integrating antennas with TEGs and solar cells are also particularly suitable for smart-fabric interactive-textile systems in a variety of applications such as rescue missions, interventions, and health care [115].

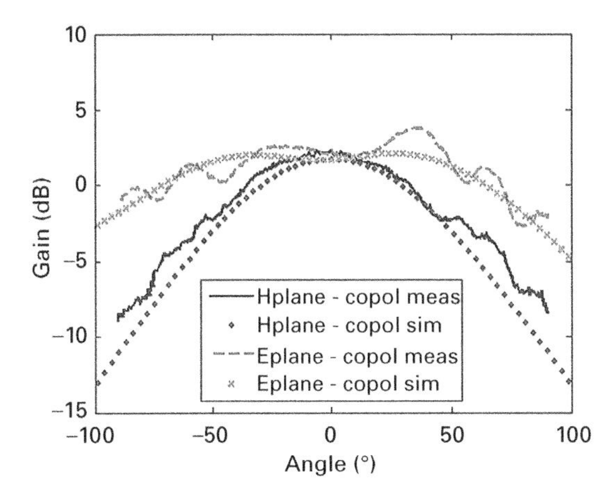

Figure 5.12 Measured gain radiation pattern of the patch antenna with TEG [114].

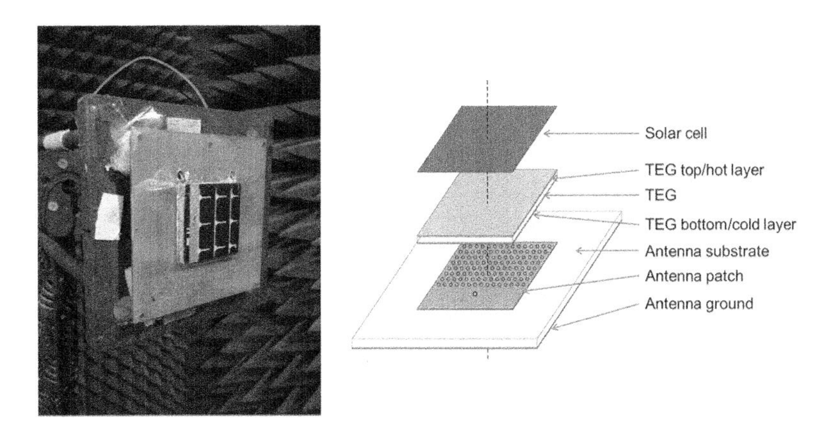

Figure 5.13 Prototype of shorted patch antenna with TEG and solar cell [114]. Antenna photo courtesy of Dr. Marco Virili, Qorvo Inc. Circuit schematic ©2015 IEEE. Reprinted with permission from [114]

5.8 Problems and Questions

1. Describe the three thermoelectric phenomena: (a) the Seebeck effect, (b) the Peltier effect, and (c) the Thomson effect.
2. Describe the three mechanisms of heat transfer.
3. Describe Fourier's law and its analogy to Ohm's law.
4. Describe the conservation of energy in the hot and cold surface boundaries of a TEG.
5. Derive the Carnot efficiency formula assuming the Peltier and Seebeck effects.
6. Calculate the emitted heat from the cold boundary surface of a TEG by solving the heat diffusion equation and applying Fourier's law and the conservation of energy.

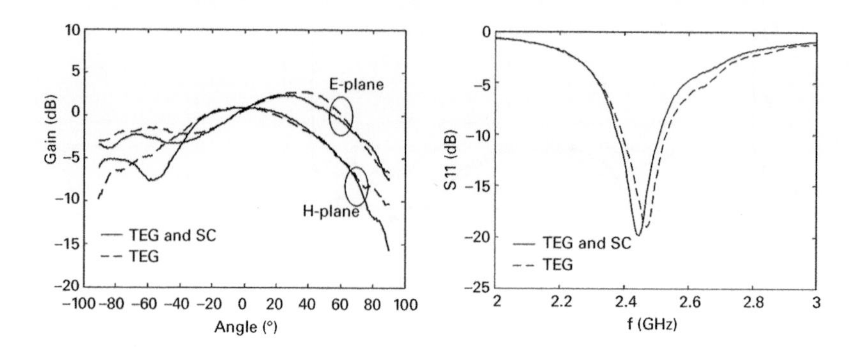

Figure 5.14 Measured performance of the patch antenna with TEG and solar cell [114].

7. Starting from (5.38) or (5.39) for the TEG efficiency, compute the optimum load that maximizes the efficiency and find the maximum efficiency.
8. The hot junction of a TEG is at $T_H = 315$ K and the cold junction at $T_C = 300$ K. The figure of merit of the thermoelectric material of the TEG is $Z = 0.003$ K^{-1}. The TEG comprises $N = 100$ pellet pairs, where each pellet pair has electrical resistivity $\rho = 25$ $\mu\Omega$m, surface $A = 1$ mm^2, and length $L = 2$ mm. Derive the Carnot efficiency and the TEG efficiency assuming that the TEG is connected to a $R_L = 50$ Ω load. What is the optimum load maximizing the efficiency and what is the optimum efficiency?
9. Derive the optimum TEG efficiency and optimum load when the length of the pellets of the TEG of the previous problem is doubled, i.e., $L = 4$ mm.

6 Wireless Power Transmission

6.1 Introduction

Wireless power transmission (WPT) refers to the concept of intentionally transferring power in a contactless manner aiming at powering a system located at a certain distance from the power source. There is a wide range of applications that require of devices and sensors operating in an autonomous manner, communicating with each other and providing us with useful information. Some of these applications include biomedical implants that require of recharging, devices placed in inaccessible locations, and the high number of sensors and devices necessary to implement the concepts of the Internet of Things and machine-to-machine communications. Wireless power transmission appears as an attractive solution to provide the required energy autonomy in these applications.

In a wireless power transmission system, a power source converts dc electrical power P_i to an RF signal P_t that is transmitted wirelessly to a target device, which receives the RF signal P_r and converts it back to dc electrical power P_L, in order to power itself. The target device harvests electromagnetic energy. The overall efficiency of such a system is defined as

$$\eta = \eta_{dcRF} \cdot \eta_{ap} \cdot \eta_{RFdc} \tag{6.1}$$

with

$$\eta_{dcRF} = \frac{P_t}{P_i} \tag{6.2}$$

$$\eta_{ap} = \frac{P_r}{P_t} \tag{6.3}$$

$$\eta_{RFdc} = \frac{P_L}{P_r}. \tag{6.4}$$

The first term η_{dcRF} is dominated by the efficiency of the amplifier stages of the power source. Continuous wave power amplifiers and oscillators can reach very high efficiencies, which, depending on the operating frequency and required output power, can vary from >95% in the low MHz range [116] to >70% in the low GHz range [117] and >60% in X-band [118] limited by device technology and parasitics. This chapter and Chapter 7 focus on the other two efficiency terms, the efficiency of the radiating apetures η_{ap} and the efficiency of the RF–dc conversion devices n_{RFdc}, the rectifiers.

These two efficiency terms are related to the classification of wireless power transmission systems in different categories in an indirect way. One may distinguish between high(er)-power wireless power transmission systems and low(er)-power RF harvesting systems. Typically, when intentional radiators are used, the RF power at the receiving devices is high. The high operating power has a strong effect on the resulting RF–dc conversion efficiency, which is also high. Such systems include high-power microwave power transmission systems such as wireless vehicle chargers and the solar power satellite concept [119]. UHF RFID systems, however, although they involve intentional RF power transmission, operate at low power levels with a sensitivity of approximately −20 dBm and for the purposes of this book are classified as RF harvesting systems, which is the focus of Chapter 7. In this book, we may loosely classify as RF harvesting systems ones where the input RF power at the terminals of the receiving antenna is approximately −20 dBm or less. In this case, as we will see in Chapter 7, the obtained RF–dc conversion efficiency η_{RFdc} is less than approximately 25%.

Wireless power transmission systems are also classified as near-field or far-field, depending on the electromagnetic field distribution of the transmitting and receiving radiators at the position of the receiving and transmitting antennas respectively, as described in the antenna literature, such as for example [120]. The field distribution has a strong effect on the operating efficiency of the radiating apertures η_{ap}. Brown [121] presents a plot of the efficiency of the radiating apertures versus a parameter τ,

$$\tau = \frac{\sqrt{A_t A_r}}{\lambda d}, \tag{6.5}$$

where A_t and A_r are the effective aperture areas of the transmitting and receiving antennas respectively, d the distance between the transmitting and receiving antennas, and λ the free space wavelength of the continuous wave transmitted RF signal. The efficiency η_{ap} has an exponential dependence on τ [122]

$$\eta_{ap} = 1 - e^{-\tau^2}, \tag{6.6}$$

shown in Figure 6.1. The well-known Friis transmission formula [120] is valid in the far-field of the radiating apertures and gives

$$P_r = P_t \frac{\lambda^2 G_t G_r}{(4\pi d)^2} = P_t \frac{A_t A_r}{(\lambda d)^2} \Rightarrow \eta_{ap} = \tau^2 \tag{6.7}$$

where the transmitting and receiving effective aperture areas are related to the transmitting and receiving antenna gain as $A_t = G_t\lambda^2/(4\pi)$ and $A_r = G_r\lambda^2/(4\pi)$ respectively. One can easily verify that the expression of the efficiency of the radiating apertures derived from the Friis transmission formula represents a first-order approximation of 6.6 for small τ. The agreement between the two expressions is good for $\tau < 0.5$, which corresponds to the Fraunhofer limit for

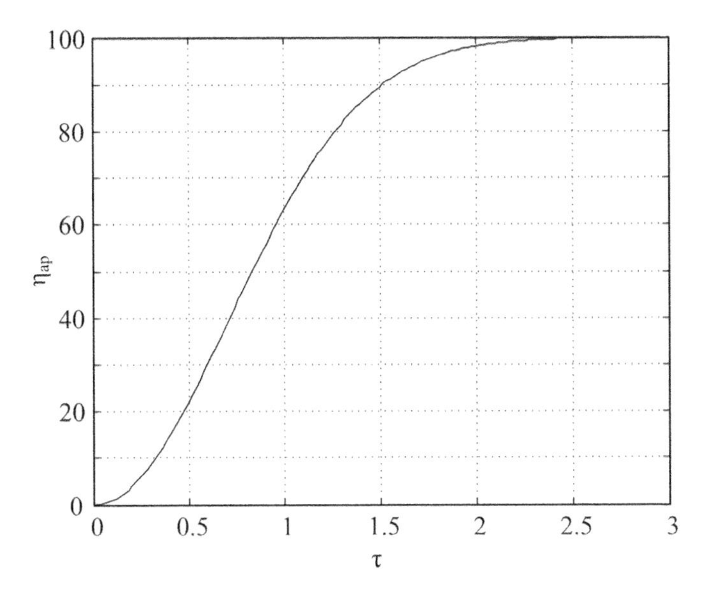

Figure 6.1 Efficiency of the radiating apertures versus τ.

the far-field [120]. As τ increases beyond 0.5, the Friis transmission formula does not apply and the system enters the near-field region. It is interesting to note that for $\tau > 2$ an efficiency of the radiating apertures of nearly 100% can be obtained.

This chapter covers near-field and far-field wireless power transmission systems, including system analysis, design guidelines, and measurement techniques. This chapter also covers, as part of near-field wireless power transmission systems, nonresonant and resonant inductive coupling covering different aspects such as impedance matching, appearing modes under strong coupling conditions, and misalignment effects. Systems employing capactive coupling are also discussed. The far-field RF/microwave radiation–based wireless power transmission section focuses on rectenna arrays for high-power transmission.

6.2 Historical Perspective

The concept of power transmission by electromagnetic waves initially appeared in the works of Hertz and Tesla [123, 124, 125]. The first wireless power transfer experiments were performed back in 1899 when Tesla tried to wirelessly transfer energy by using large coils [124, 125] at 150 KHz. Later on, W. C. Brown proposed wireless power transmission making use of higher frequencies such as microwaves in order to achieve further transmission distances [119, 126]. Toward this objective, he developed the rectenna element, comprising an antenna and a rectifier circuit connected to its terminals, which he patented in 1969 [127].

Based on the works of Brown, the first applications of wireless power transmission focused on directive high-power transmission, such as the works on solar power satellites, appeared [119, 128, 129, 130]. Solar energy was captured and converted to electromagnetic signals that then could be reradiated and used to power devices at long distances. Recent interest in compact devices and sensors with energy harvesting capability has led some to utilize the same principle used in solar power satellites toward powering low-power electronics and sensors. As a result, low-power but highly efficient solar-to-RF converters–based solar active antennas using dc-to-RF conversion circuits such as class-E oscillators have been considered [77, 131].

Since the appearance of RF/microwave wireless power transmission, there have been a large number of works toward maximizing the power transfer efficiency in these systems mainly focusing on maximizing the RF-to-dc conversion efficiency of rectifier circuits and rectenna elements [12, 122, 132, 133, 134, 135, 136, 137, 138, 139, 140, 141, 142], synthesizing rectenna arrays for high-power wireless power transmissions [143, 144, 145, 146] and more lately there have been several efforts on the optimal transmitting signal waveform design [147, 148, 149].

On the other hand, near-field inductive coupling has also been widely studied [150, 151, 152, 153, 154, 155, 156], showing good performance from distances in the order of a few millimeters up to a couple of meters. Several methods to analyze and optimize the performance of inductive coupling systems have been used such as coupled mode theory (CMT) [150, 151, 152] and coupled inductance model circuit theory [153, 154]. CMT focuses on representing the inducting coupling system using the theory behind coupled resonators and represents the system using first-order differential equations. Circuit theory derives equations similar to the ones used in transformer circuits to describe the behavior of inductive coupling systems.

Nonresonant inductive coupling wireless power transmission has been widely used for short distances in the millimeter range and up to a few centimeters. As the targeted powering distance increases, nonresonant inductive coupling can no longer be used. Resonant inductive coupling where the transmitting and receiving coils are made to resonate by introducing additional capacitors with the adequate values allows maximizing the power transfer efficiency for the selected operating frequency. Resonant inductive coupling has been shown to achieve ranges of up to a few meters [157].

The wireless transmission of power to power up devices is pending regulation and standardization. The existing Qi standard was developed by the Wireless Power Consortium [158] and focuses on low-power (up to 5 W) inductive coupling within distances up to 40 mm. Further efforts are going on toward extending this standard to include medium power levels up to 120 W.

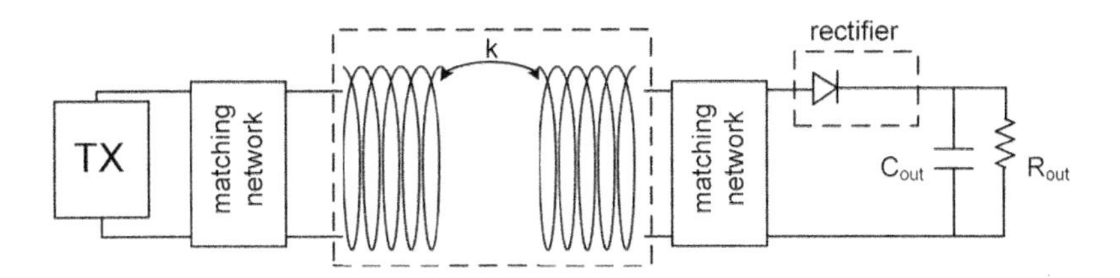

Figure 6.2 Basic building blocks of an inductive coupling wireless power transfer system.

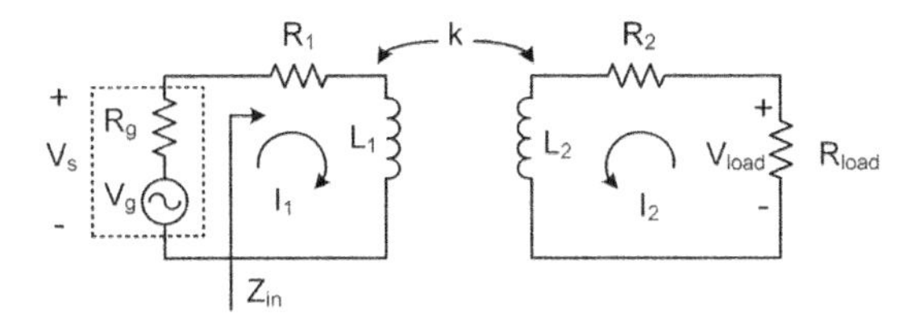

Figure 6.3 Circuit model of a nonresonant inductive coupling system.

6.3 Near-Field Wireless Power Transmission

An inductive coupling system comprises a transmitter, two inductors between which the transfer of power occurs, a rectifier circuit that converts the received power to dc and a load (Figure 6.2). Additionally, matching networks are required in the transmitting and receiving sides in order to maximize the transfer of power between the transmitter and the transmission inductor and between the receiving inductor and the rectifier circuit respectively.

6.3.1 Nonresonant Inductive Coupling

Nonresonant inductive coupling refers to the method of wirelessly transferring power by using two coupled inductors. Its working principle is based on the fact that an existing current in the transmitting coil L_1 generates a magnetic field. If the two inductors are close enough, then a change in the magnetic flux in L_1 induces a current in the receiving coil L_2. The power transfer efficiency in these nonresonant inductive coupling systems depends on several factors such as the coupling coefficient, the size of the coils, and their geometry and the alignment between transmitting and receiving coils. In order to analyze the behavior of these systems, circuit theory is used to model the power transfer process.

Figure 6.3 shows the schematic of a basic representation of a nonresonant inductive coupling system where L_1 and L_2 are the two coupled coils and R_1 and R_2 represents the internal resistances associated to L_1 and L_2.

Applying Kirchoffs law, the equations corresponding to the circuit in Figure 6.3 can be written as follows:

$$\begin{aligned} V_s &= (R_1 + j\omega L_1)I_1 + j\omega M I_2 \\ 0 &= j\omega M I_1 + (R_2 + R_{load} + j\omega L_2)I_2. \end{aligned} \tag{6.8}$$

M is the mutual inductance among the two inductors, and it is defined as

$$M = k\sqrt{L_1 L_2}, \tag{6.9}$$

where k is the coupling coefficient among the inductors. Depending on the value of k, different types of coupling can be considered among the inductors, with $k < 0.5$ corresponding to weak coupling and $k > 0.5$ to strong coupling.

The coupling coefficient k depends of the distance between the two coils. The farther the transmitting and receiving coils are placed, the lower the value of k. In the same manner, the mutual inductance M depends on several parameters, such as the dimensions of the coils, the distance among them, and their relative position [159, 160, 161].

From (6.8), the power transfer efficiency η in a nonresonant inductive coupling system can be calculated as

$$\eta = \frac{P_{load}}{P_S} \tag{6.10}$$

with

$$\begin{aligned} P_S &= \frac{V_s I_1^*}{2} \\ P_{load} &= \frac{V_{load} I_2^*}{2} \\ V_{load} &= -I_2 R_{load}, \end{aligned} \tag{6.11}$$

where P_{load} is the average delivered power to the load, P_S is the average input power delivered at Z_{in} by the source, and the $()*$ operator indicates complex conjugate.

Using (6.8) through (6.11), one obtains

$$V_{load} = \frac{-j\omega M V_S R_{load}}{(R_2 + R_{load} + j\omega L_2)(R_1 + j\omega L_1) + \omega^2 M^2} \tag{6.12}$$

and finally

$$\eta = \frac{R_{load}}{R_2 + R_{load}} \frac{\omega^2 k^2 L_1 L_2}{[(R_2 + R_{load})^2 + (\omega L_2)^2] \frac{R_1}{R_2 + R_{load}} + \omega^2 k^2 L_1 L_2}. \tag{6.13}$$

As an example, consider a nonresonant inductive coupling circuit with $L_1 = L_2 = 10$ nH, $R_1 = R_2 = 2\ \Omega$ designed to operate at 13.56 MHz, shown in Figure 6.4. The performance of the circuit shown in Figure 6.4 in terms of power transfer efficiency can be evaluated using the preceding formulas or using a commercial simulator. Figure 6.5 shows the efficiency versus frequency for different values of the coupling coefficient k. It can be observed that this system requires a high value of k to achieve good power transfer efficiencies. As k is

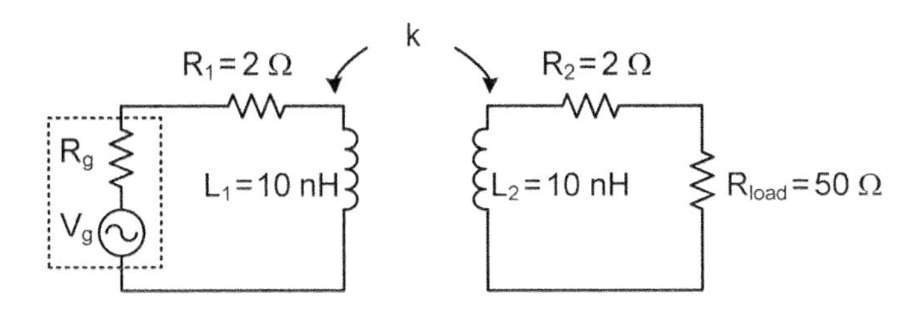

Figure 6.4 Nonresonant inductive coupling system.

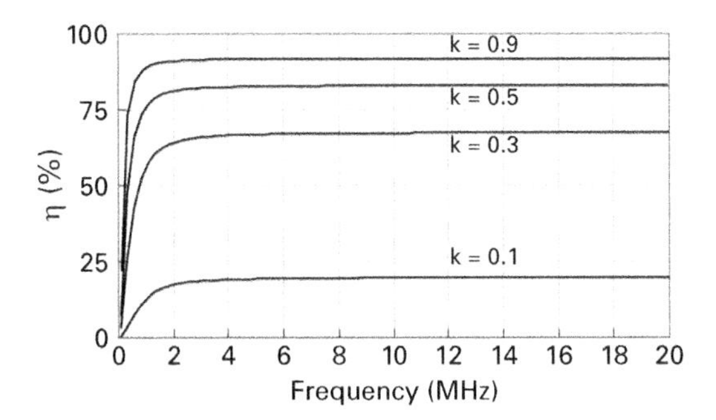

Figure 6.5 Power transfer efficiency of the nonresonant inductive coupling system of Figure 6.4.

related to the distance among the coils, this implies that the coils need to be relatively close to achieve good power transfer efficiency. This fast degradation of the power transfer efficiency with distance is the main limitation in nonresonant inductive coupling systems.

The power transfer efficiency in these systems also has a strong dependence on the value of R_{load} connected at the output of the circuit. There exists an optimum value of the load for which the power transfer efficiency is maximum. Figure 6.6 shows the power transfer efficiency versus R_{load} of the circuit of Figure 6.4 for a fixed coupling coefficient value $k = 0.1$. This result shows that the initially selected $R_{load} = 50\ \Omega$ does not lead to maximum efficiency.

6.3.2 Resonant Inductive Coupling

As stated previously, the power transfer efficiency in nonresonant inductive coupling systems degrades rapidly as the distance between the transmitting and receiving ends increases. One way to eliminate this limitation in the transfer distance is to consider resonant inductive coupling [150, 151, 152, 153, 154]. In resonant inductive coupling systems, the inductances L_1 and L_2 are made to resonate with two capacitors C_1 and C_2 introduced in the system such that both

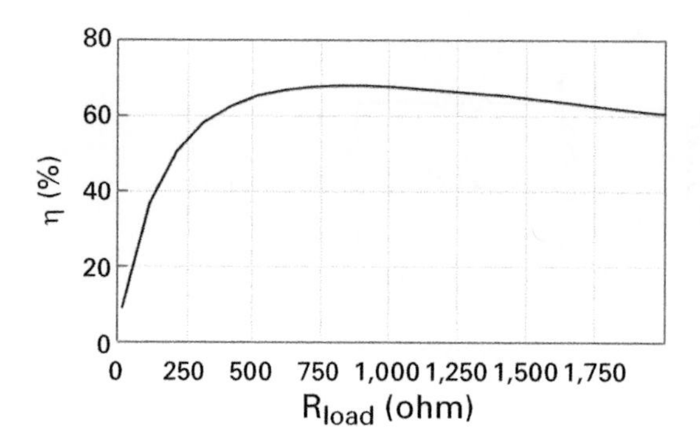

Figure 6.6 Power transfer efficiency of the nonresonant inductive coupling system in Figure 6.4 versus the output load R_{load} for a coupling coefficient of $k = 0.1$.

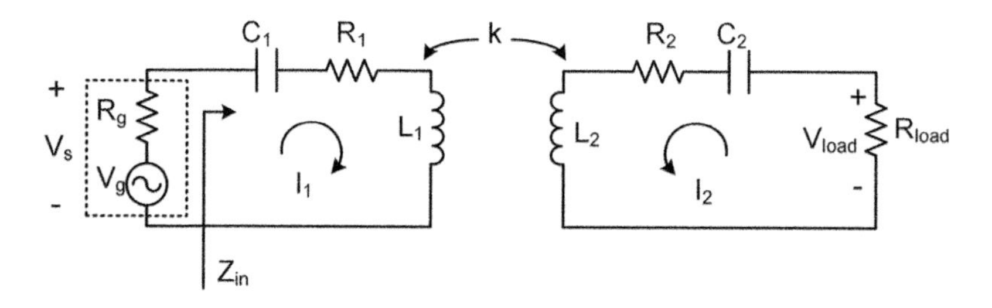

Figure 6.7 Schematic of a series–series resonant inductive coupling system.

pairs (L_1, C_1) and (L_2, C_2) resonate at the same frequency (Figure 6.7). The resonance frequencies of both resonators can be selected as

$$\omega_o = \frac{1}{\sqrt{L_1 C_1}} = \frac{1}{\sqrt{L_2 C_2}}. \tag{6.14}$$

There are four possible topologies that can be used to tune out the primary and secondary inductances, namely series–series, series–shunt, shunt–series, and shunt–shunt, shown in Figure 6.8.

The series–series topology leads to a simple mathematical analysis and will be considered next. Applying Kirchoffs law, the equations of the resonant inductive coupling system can be written as in (6.15):

$$\begin{aligned} V_s &= (R_1 + j\omega L_1 + \frac{1}{j\omega C_1})I_1 + j\omega M I_2 \\ 0 &= j\omega M I_1 + (R_2 + R_{load} + j\omega L_2 + \frac{1}{j\omega C_1})I_2. \end{aligned} \tag{6.15}$$

Using (6.11) and (6.15), the power transfer efficiency becomes as follows

$$\eta = \frac{R_{load}}{R_2 + R_{load}} \frac{\omega^2 k^2 L_1 L_2}{\left[(R_2 + R_{load})^2 + (\omega L_2 - \frac{1}{\omega C_2})^2\right] \frac{R_1}{R_2 + R_{load}} + \omega^2 k^2 L_1 L_2}. \tag{6.16}$$

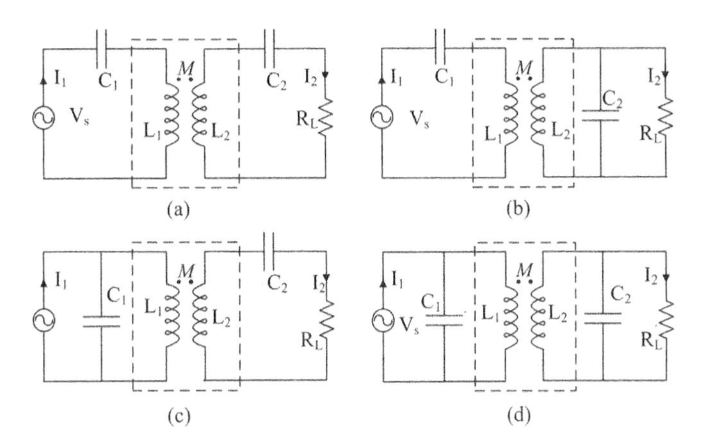

Figure 6.8 Different topologies of resonant inductive coupling systems: (a) series–series, (b) series–shunt, (c) shunt–series, and (d) shunt–shunt.

At the resonance frequency, the power transfer efficiency is maximum and can be expressed as

$$\eta_o = \frac{R_{load}}{R_2 + R_{load}} \frac{\omega^2 k^2 L_1 L_2}{R_1(R_2 + R_{load}) + \omega^2 k^2 L_1 L_2}. \tag{6.17}$$

In the same way as it happens with the nonresonant inductive coupling systems, the resonant inductive coupling system presents maximum power transfer efficiency for a certain value of R_{load} that has to be carefully selected in order to optimize the efficiency of the system.

The power transfer efficiency can also be expressed in terms of the quality factors of the inductances L_1 and L_2. Considering that the loaded quality factors of the inductances are $Q_1 = \omega L_1/R_1$ and $Q_2 = \omega L_2/R_2$ and that the quality factor of the load can be expressed as $Q_{load} = \omega L_2/R_{load}$, the power transfer efficiency expression reduces to

$$\eta_o = \frac{k^2 Q_1 Q_L}{1 + k^2 Q_1 Q_L} \frac{Q_L}{Q_{load}}, \tag{6.18}$$

where

$$Q_L = \frac{Q_{load} Q_2}{Q_{load} + Q_2} \tag{6.19}$$

is the loaded quality factor [45].

As an example, consider a resonant inductive coupling circuit with $L_1 = L_2 = 10$ nH, $R_1 = R_2 = 2\ \Omega$ designed to operate at 13.56 MHz (Figure 6.9). Capacitances C_1 and C_2 are introduced in the circuit of Figure 6.4 in order to resonate with L_1 and L_2 at 13.56 MHz. The capacitances are calculated using the expression for the resonant frequency of a resonator. In Figure 6.10, it can be seen that the power transfer efficiency of the system remains at a high value for the resonant frequency of 13.56 MHz independently of the coupling coefficient value. This shows that using a resonant inductive coupling system, it is possible

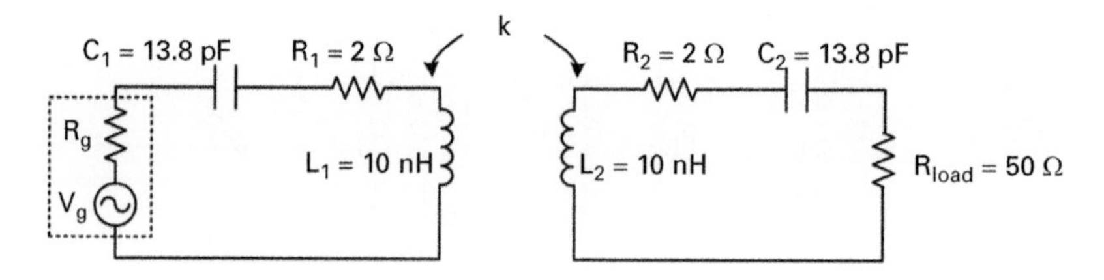

Figure 6.9 Resonant inductive coupling system example.

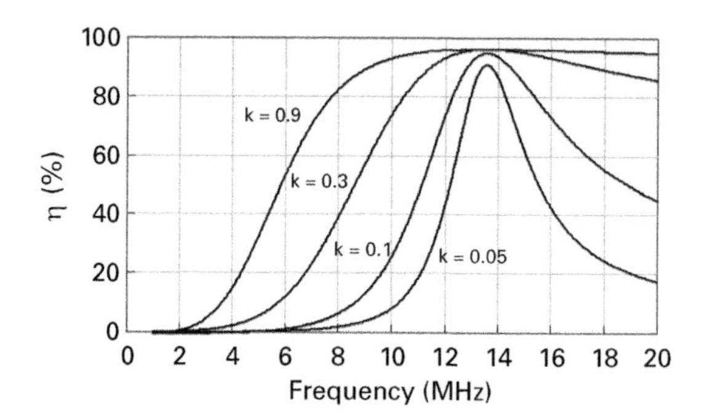

Figure 6.10 Power transfer efficiency versus frequency for different values of the coupling coefficient k.

to obtain high values of power transfer efficiency for higher distances than when using a nonresonant inductive coupling system.

6.3.3 Strong Coupling in Resonant Inductive Coupling Systems

In inductive coupling systems, additionally to the power transfer efficiency, another parameter that has to be optimized is the amount of power delivered to the load as this will determine how much dc power can be obtained when using these systems together with the RF-to-dc converter. As it can be expected, the amount of power delivered to the load depends directly on the coupling coefficient k. It could be expected that the higher the coupling coefficient k, the higher the amount of power that will be efficiently delivered to the load. This directly applies in nonresonant inductive coupling systems; however, this is not the case in resonant inductive coupling systems. Due to the coupled resonators theory, as k increases the coupling between the two inductors gets stronger and several operation modes may appear in the system. These modes produce a double peak in the output power curves around the resonance frequency, indicating that the maximum delivered power does not occur at the desired resonance frequency [154].

The location of the peaks in the output power versus frequency can be determined by calculating the derivative of P_{load} with respect to ω and setting it equal to zero as

$$\frac{\partial P_{load}}{\partial \omega} = 0. \tag{6.20}$$

Potential ways to correct this frequency splitting effect include varying the values of C_1 and C_2 in the transmitting and the receiving ends or modifying the input/output matching networks as a manner to shift one of the peaks so that they are centered at the desired operation frequency [162].

As an example, consider the resonant inductive coupling circuit of Figure 6.9. Figure 6.11 shows the output power versus frequency for different coupling coefficient values for the circuit of Figure 6.9 that is designed to operate at 13.56 MHz. For low levels of coupling among the inductors, there is a single peak in the curve that is centered in the resonance frequency of 13.56 MHz. The output power curve begins presenting a double peak for coupling coefficient values above $k = 0.1$ (Figure 6.11b). As k continues to increase the valley between the two peaks is more pronounced and the peaks are more spaced (Figure 6.11c). When this phenomenon begins, the resonant frequency of 13.56 MHz falls in the valley between the two peaks, which means the delivered power to the load is reduced.

The second peak can be shifted back to 13.56 MHz by tuning the value of $C_1 = C_2 = C$ from the initial value of 13.8 pF to a new value of 19.5 pF. The results obtained after applying this tuning can be seen in Figure 6.12. In a similar manner, the first peak could be shifted to 13.56 MHz by tuning the value of C in the opposite direction. The applied correction may lead to reduced power transfer efficiency between the inductors, but it will improve the global performance of the system when connected to the RF-to-dc converter.

6.3.4 Impedance Matching in Inductive Coupling Systems

The power transfer efficiency calculated using (6.10) refers to the ratio between the delivered power to the load over the delivered power to Z_{in} at the input of the system of Figures 6.3 or 6.7.

The calculation of the power transfer efficiency referred to the available power at the transmitting source P_a has to take into account the mismatch between the source impedance R_g and the input impedance of the first coil Z_{in}. This efficiency can be expressed as follows

$$\eta_{av} = \frac{P_{load}}{P_a}. \tag{6.21}$$

The relationship between the delivered power P_s to Z_{in} and the available power in the source P_a can be written as

$$P_s = P_a\left[1 - |\Gamma_{in}|^2\right], \tag{6.22}$$

where Γ_{in} is the input reflection coefficient $\Gamma_{in} = (Z_{in} - R_g)/(Z_{in} + R_g)$. $Z_{in} = R_{in} + jX_{in}$ is the complex input impedance of the system. We have assumed

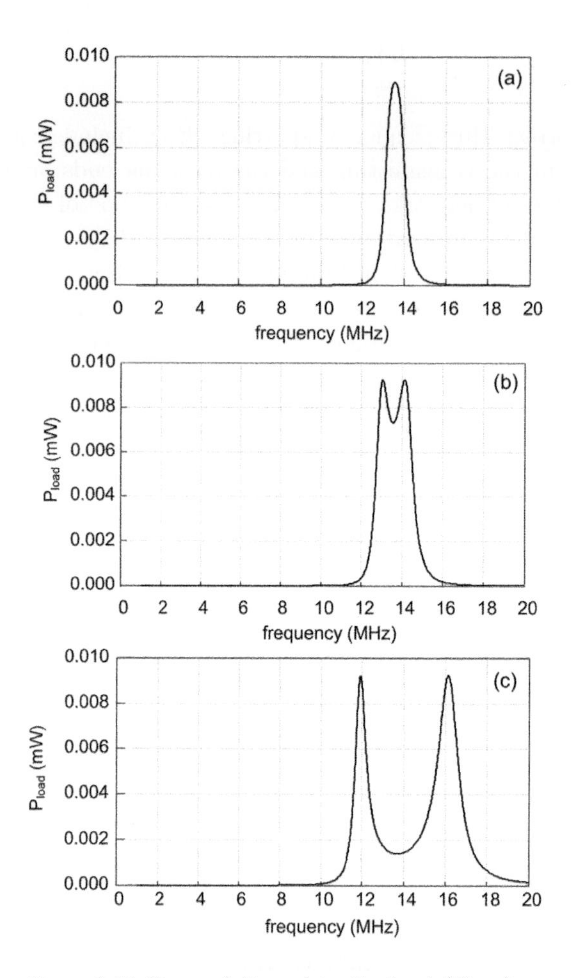

Figure 6.11 Power delivered to the load (P_{load}) versus frequency for different coupling coefficient k values (a) $k = 0.05$ (b) $k = 0.1$ (c) $k = 0.3$.

without loss of generality that the source has a real impedance R_g. Substituting (6.22) into (6.21),

$$\eta_{av} = \eta \left[1 - |\Gamma_{in}|^2\right]. \tag{6.23}$$

From (6.22), it is inferred that the maximum power transfer occurs when $|\Gamma_{in}|$ is minimum. It is straightforward to show that this occurs when $R_g = R_{in}$ and it equals $\eta_{av} = \eta$. The available maximum efficiency η_{av} represents a lower bound of the maximum efficiency η of the system.

In order to maximize the power transfer efficiency, it is important that both the source and the load are matched to the inductive coupling system. There are several ways to implement impedance matching, for example using reactive networks comprising series and parallel capacitances and inductances in different configurations in order to achieve $Z_{in} = R_g$.

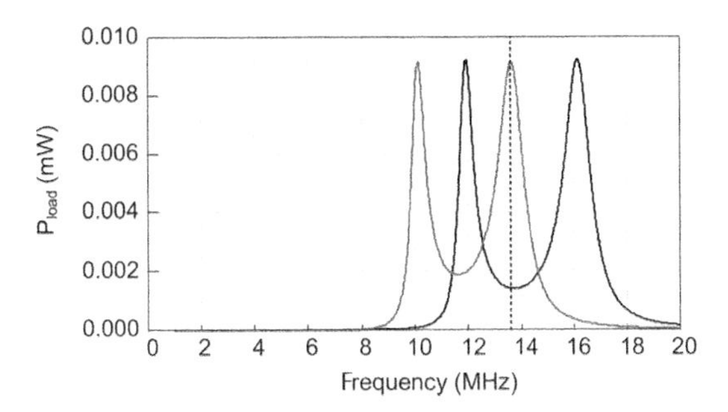

Figure 6.12 Power delivered to the load (P_{load}) versus frequency for $k = 0.3$ corrected to reach its peak value at 13.56 MHz.

6.3.5 Misalignment Effects

In the same manner as the value of the coupling coefficient k varies with distance, it is also affected by misalignments between the transmitting and receiving coils. The proper alignment of the coils in an inductive coupling system is a key parameter as the power transfer efficiency may decrease dramatically with misalignments.

Different types of misalignments may occur in an inductive coupling system (Figure 6.13):

1. Lateral misalignment: The transmitting and receiving inductors are in parallel planes but their centers are not aligned but displaced laterally a distance Δ.
2. Angular misalignment: The plane of the receiving loop is rotated by an angle θ with respect to the plane of the transmitting loop.
3. Lateral/angular misalignments: Both lateral and angular misalignments occur simultaneously.

The variation in the coupling coefficient k is due to the change in the mutual inductance between the two inductors depending on the alignment conditions. In order to estimate the variation in k, it is necessary to calculate the mutual inductance for the different alignment cases. The usual manner to solve for the mutual inductance is to use the current filament method [163, 164] that divides the cross section of the coils as a mesh of (2M + 1) by (2N + 1) for L_1 and a mesh of (2m + 1) by (2n + 1) for L_2. The mutual inductance using the filament method can be expressed as

$$M_{12} = \frac{N_1 N_2}{(2M+1)(2N+1)(2m+1)(2n+1)} \sum_j \left(\sum_i M_{ij} \right). \tag{6.24}$$

The mutual inductances M_{ij} can also be calculated using Newmanns form

$$M_{12} = \frac{\mu}{4\pi} \oint \oint \frac{dl_1 dl_2}{r_{12}}, \tag{6.25}$$

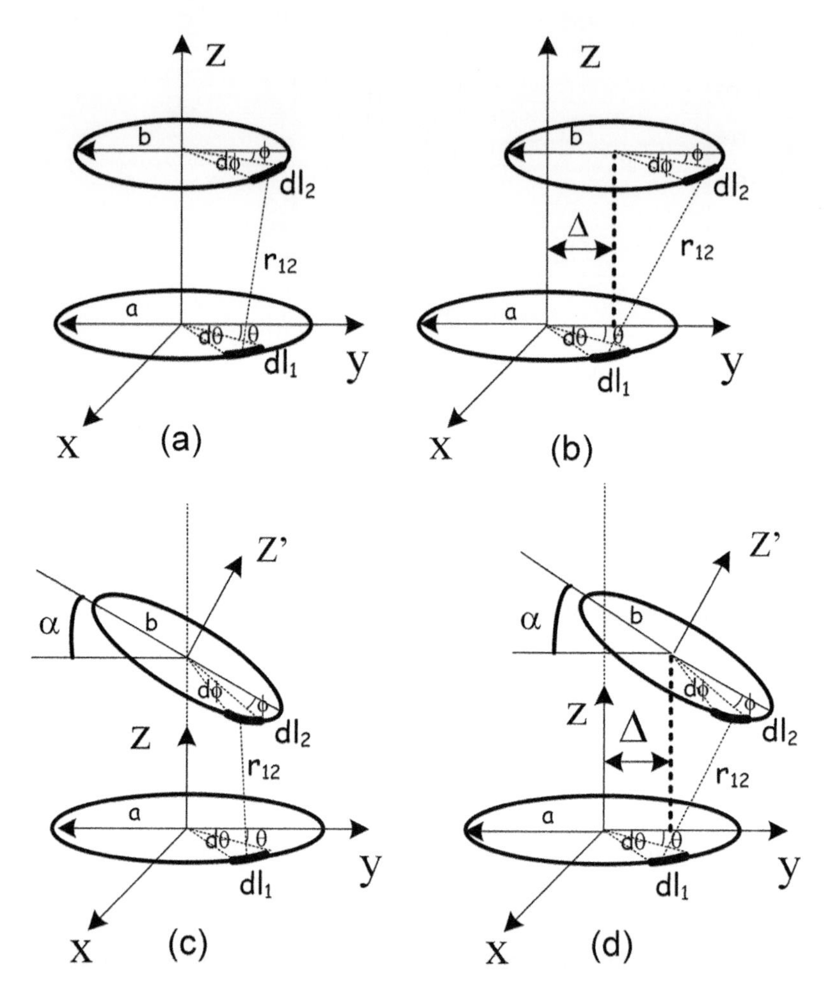

Figure 6.13 Different types of misalignments between the transmitting and receiving inductors in an inductive coupling wireless power transfer system (a) perfect alignment, (b) lateral misalignment, (c) angular misalignment, and (d) lateral and angular misalignment.

where r_{12} is the distance between the element dl_1 of coil L_1 and the element dl_2 of coil L_2.

For the aligned coil system, the distance between dl_1 and dl_2 is

$$r_{12} = \sqrt{\alpha^2 + b^2 - 2\alpha b\cos(\theta - \phi) + d^2}, \tag{6.26}$$

which results in

$$M_{12} = \mu\sqrt{\alpha b}\left[\left(\frac{2}{x} - x\right)K(x) - \frac{2}{x}E(x)\right] \tag{6.27}$$

with

$$x^2 = \frac{4\alpha b}{(\alpha + b)^2 + d^2}. \tag{6.28}$$

$K(x)$ and $E(x)$ are the complete elliptic integrals of the first and second kind respectively that are defined as

$$K(x) = \int_0^{\pi/2} \frac{d\phi}{\sqrt{1 - x^2 \sin^2(\phi)}}$$
$$E(x) = \int_0^{\pi/2} \sqrt{1 - x^2 \sin^2(\phi)} d\phi. \tag{6.29}$$

In the case of the laterally misaligned coil system, the distance between dl_1 and dl_2 becomes

$$r_{12} = \sqrt{a^2 + b^2 + 2ab\cos(\theta - \phi) + 2(b - a)\Delta\cos(\theta) + d^2}. \tag{6.30}$$

In the same way, the distance for the case of angular misalignment can be expressed as

$$r_{12} = \sqrt{a^2 + b^2 - 2ab(\cos(\theta)\cos(\phi)\cos(\alpha) + \sin(\theta)\sin(\phi)) - 2bd\cos(\phi)\sin(\alpha) + d^2}. \tag{6.31}$$

6.3.6 Measurements in Inductive Coupling Systems

There are several key parameters that need to be accurately measured when designing inductive coupling systems for wireless power transmission. The main and most common parameters include self-inductance of the inductors, mutual inductance between the inductors, quality factor of resonators when designing resonant inductive coupling systems, and power transfer efficiency. Some basic measurements guidelines are given next in order to determine these parameters.

The self-inductances L_1 and L_2 and the mutual inductance M of two coupled inductors can be obtained by measuring the impedance matrix $[Z]$ of the structure. This measurement can be done by using a conventional vector network analyzer (VNA) that allows obtaining $[Z]$ directly or by measuring the scattering parameter matrix $[S]$ and then calculating the impedance matrix $[Z]$ by using the well-known corresponding transforming relationships between $[Z]$ and $[S]$ [165].

Considering the circuit equations for a two coupled inductors structure and rearranging them in a matrix form as in (6.32) it is possible to determine L_1, L_2, and M using (6.33) through (6.36). The internal resistance of the inductors has been considered in the equations of the system (6.32).

$$\begin{bmatrix} V_1 \\ V_2 \end{bmatrix} = \begin{bmatrix} R_1 + j\omega L_1 & j\omega M \\ j\omega M & R_2 + j\omega L_2 \end{bmatrix} \cdot \begin{bmatrix} I_1 \\ I_2 \end{bmatrix} \tag{6.32}$$

$$\begin{bmatrix} Z_{11} & Z_{12} \\ Z_{21} & Z_{22} \end{bmatrix} = \begin{bmatrix} R_1 + j\omega L_1 & j\omega M \\ j\omega M. & R_2 + j\omega L_2 \end{bmatrix} \tag{6.33}$$

$$M_{12} = M_{21} = M = \frac{\Im(Z_{12})}{\omega} = \frac{\Im(Z_{21})}{\omega} \tag{6.34}$$

$$L_1 = \frac{\Im(Z_{11})}{\omega} \tag{6.35}$$

$$L_2 = \frac{\Im(Z_{22})}{\omega}, \tag{6.36}$$

where the function $\Im()$ denotes the imaginary part of its argument.

Another parameter that needs to be determined when designing resonant inductive coupling systems is the quality factor of the resonators formed by (L_1, C_1) and (L_2, C_2) as these quality factors affect the power transfer efficiency of the system. There are several methods to measure the quality factor of resonator circuits [166, 167], such as the reflection method and the transmission method.

A way to measure the quality factor in a simple manner using scattering parameters measurement follows. The loaded quality factor can be extracted from the magnitude of S_{21} with the help of (6.37):

$$Q_L = \frac{f_o}{\Delta f}, \tag{6.37}$$

where f_o is the resonance frequency and Δf is the bandwidth at which the magnitude of S_{21} falls to half of its value or equivalently falls by 3 dB with respect to its peak value.

If it is assumed that the coupling between the resonator and the input and output networks used in the measurements setup is weak, then the following approximation for calculating the unloaded quality factor holds:

$$Q_U = \frac{Q_U}{1 - |S_{21}|^2}. \tag{6.38}$$

If the coupling of the resonator with the input and output networks is strong, then more complex methods need to be used [166, 167].

The power transfer efficiency in an inductive coupling system can also be determined by using scattering parameter measurements. The efficiency can be expressed in terms of the scattering parameters as

$$\eta = \frac{P_{load}}{P_S} = \frac{|S_{21}|^2(1 - |\Gamma_{load}|^2)}{(1 - |\Gamma_{in}|^2)|1 - S_{22}\Gamma_{load}|^2}, \tag{6.39}$$

where

$$\Gamma_{in} = \frac{S_{11} + S_{12}S_{21}\Gamma_{load}}{1 - S_{22}\Gamma_{load}}. \tag{6.40}$$

If both the input and output of the inductive coupling system are matched to the source resistance R_g and load resistance R_{load}, then $\Gamma_{in} = \Gamma_{load} = 0$ and the power transfer efficiency equation is reduced to

$$\eta = |S_{21}|^2. \tag{6.41}$$

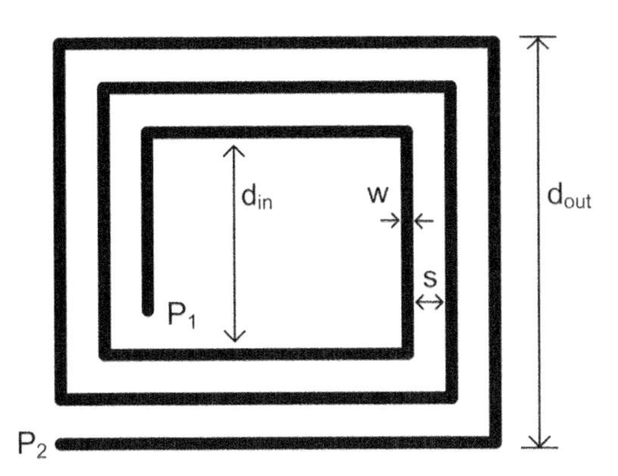

Figure 6.14 Printed inductor designed using the modified Wheeler formula.

For example, consider the design of a printed inductor on Arlon 25N substrate with $N = 3$, $w = 0.25$ mm and $s = 0.5$ mm. Depending on the geometry of the inductors that will be used for a specific design of wireless power transfer using inductive coupling, the initial dimensions of the inductors can be calculated using existing formulas and expressions [159, 160, 161, 168].

The rectangular printed inductor in Figure 6.14 has been designed using the modified Wheeler formula [160, 161], shown in (6.42), to present approximately an inductance value of $L = 0.1$ μH,

$$L = K_1 \mu_o \frac{N^2 d_m}{1 + K_2 \phi}, \tag{6.42}$$

where $\mu_o = 4\pi\dot{1}0^{-7}$ Hm^{-1} is the vacuum permeability, N is the number of turns of the inductor, K_1 and K_2 are layout-dependent parameters, and d_{in} and d_{out} refer to the inner and outer dimensions of the inductor (Figure 6.14). Finally, d_m and ϕ are defined as

$$d_m = \frac{d_{in} + d_{out}}{2} \tag{6.43}$$

$$\phi = \frac{d_{in} - d_{out}}{d_{in} + d_{out}}. \tag{6.44}$$

Assuming it is necessary to measure the inductance value of the inductor in Figure 6.14, it is possible to do so using a VNA and measuring its scattering parameters. The two ports of the inductor are marked as P_1 and P_2 in Figure 6.14. In order to measure the inductance, port P_1 is grounded and the Z-parameters are measured at P_2. If it is not possible to measure directly the Z-parameters, then one can measure Z-parameters and use the corresponding transformation relationships. The value of L can be obtained using

$$L = \frac{\Im(Z_{11})}{\omega}. \tag{6.45}$$

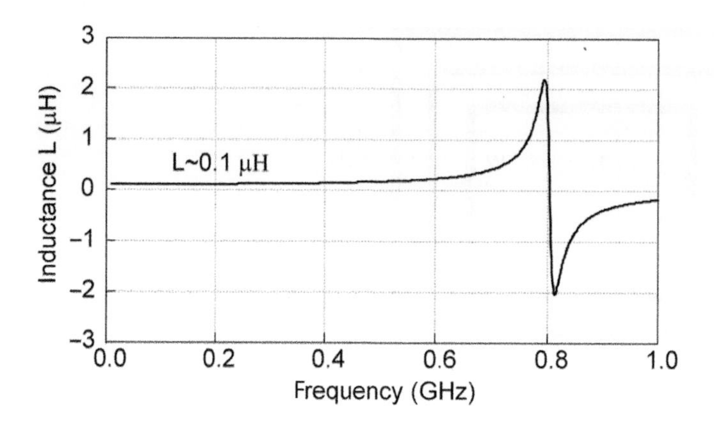

Figure 6.15 Inductance calculation from Z-parameters.

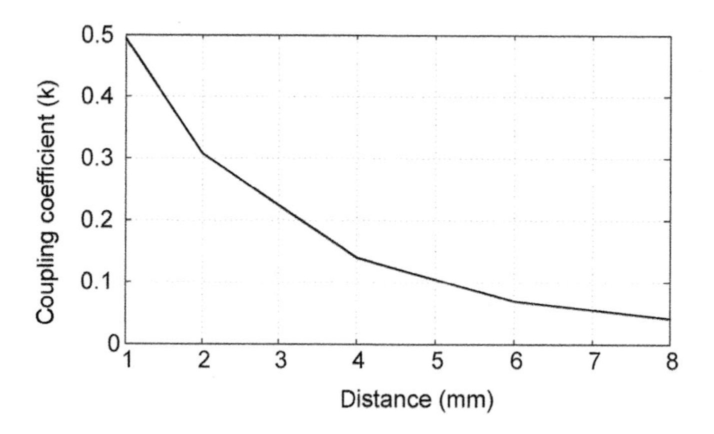

Figure 6.16 Coupling coefficient k variation with the spacing between two printed inductors as the one in Figure 6.14.

The inductance value from the printed inductor in Figure 6.14 calculated from the Z-parameters is shown in Figure 6.15. It can be observed that the obtained value is approximately $L = 0.1\ \mu$H as it was expected. In Figure 6.15, one can observe a resonance appearing around 0.8 GHz. This self-resonance of the inductor is a consequence of its own internal capacitance.

The variation of the coupling coefficient k in a two coupled inductor system can be measured by varying the distance between the two inductors and calculating k from the measured mutual inductance M and the self-inductances L_1 and L_2 obtained from (6.34) through (6.36) using the relationship $M = k\sqrt{L_1 L_2}$. The coupling coefficient k variation versus the spacing between two inductors like the one in Figure 6.14 is shown in Figure 6.16. As expected, k decreases as the distance between inductors increases.

6.3.7 Multiresonator Systems

As we have seen, there exist different methods to analyze and optimize the performance of inductive coupling systems, such as CMT [150, 151, 152] and coupled inductance circuit theory [153, 154]. CMT focuses on representing the inducting coupling system using the theory behind weakly coupled resonators and represents the system using first-order differential equations [152]. Circuit theory applies Kirchhoff's circuit laws to analyze the circuit behavior. In this section, we revisit the analysis of the steady state of near-field inductive coupling systems in order to determine the different operating modes, similarly to the approach followed in [169].

Let's consider first the basic resonant inductive coupling system shown in Figure 6.7. Applying Kirchoffs law and using $\omega_o = 1/\sqrt{L_1C_1} = 1/\sqrt{L_2C_2}$, the two equations representing the steady state of the system become

$$\begin{bmatrix} V_g \\ 0 \end{bmatrix} - \begin{bmatrix} R_g & 0 \\ 0 & R_L \end{bmatrix} \begin{bmatrix} I_1 \\ I_2 \end{bmatrix} = [Z] \cdot \begin{bmatrix} I_1 \\ I_2 \end{bmatrix}, \tag{6.46}$$

where

$$Z = \begin{bmatrix} R_1 + j\omega L_1 \left(1 - \frac{\omega_o^2}{\omega^2}\right) & j\omega M \\ j\omega M & R_2 + j\omega L_2 \left(1 - \frac{\omega_o^2}{\omega^2}\right) \end{bmatrix}. \tag{6.47}$$

One can rearrange (6.46) moving all resistance terms on the left-hand side resulting in

$$\begin{bmatrix} V_g \\ 0 \end{bmatrix} - \begin{bmatrix} R_g + R_1 & 0 \\ 0 & R_L + R_2 \end{bmatrix} \begin{bmatrix} I_1 \\ I_2 \end{bmatrix} = j\omega\,[X] \cdot \begin{bmatrix} I_1 \\ I_2 \end{bmatrix}, \tag{6.48}$$

where

$$X = \begin{bmatrix} L_1 & M \\ M & L_2 \end{bmatrix} - \begin{bmatrix} L_1\frac{\omega_o^2}{\omega^2} & 0 \\ 0 & L_2\frac{\omega_o^2}{\omega^2} \end{bmatrix} = \begin{bmatrix} L_1 & M \\ M & L_2 \end{bmatrix} - \begin{bmatrix} L_1\lambda & 0 \\ 0 & L_2\lambda \end{bmatrix} \tag{6.49}$$

and $\lambda = \omega_o^2/\omega^2$. If the input generator V_g and the losses of the system are set to zero, $j\omega X \cdot I = 0$ is a generalized eigenvalue equation defining the natural modes of the coupled inductor system. The characteristic equation is given by

$$\det X = 0 \Rightarrow \lambda^2 - 2\lambda + \left(1 - \frac{M^2}{L_1L_2}\right) = 0. \tag{6.50}$$

We have seen, however, that $M = k\sqrt{L_1L_2}$ resulting in

$$\lambda^2 - 2\lambda + \left(1 - k^2\right) = 0. \tag{6.51}$$

The two solutions of (6.51) define the two eigenvalues of the natural modes, which are equal to

$$\lambda = 1 \pm k \Rightarrow \left(\frac{\omega}{\omega_o}\right)^2 = \frac{1}{1 \pm k}. \tag{6.52}$$

The corresponding eigenvectors become

$$\begin{bmatrix} I_1 \\ I_2 \end{bmatrix} = \sqrt{\frac{L_2}{L_1}} \begin{bmatrix} 1 \\ \pm 1 \end{bmatrix}. \tag{6.53}$$

The preceding analysis provides a mathematical description of the phenomena that was observed in Figure 6.11, where we have seen that as the coupling factor k increases the two natural modes separate in frequency. Furthermore, the currents in the two coils flow in the same or opposite directions depending on the mode. In resonant inductive coupling schemes, it is common to use multiloop structures comprising intermediate relay coils. Systems with three or four coils are commonly found in the literature [150, 153, 154, 169].

It is interesting to investigate the modes of a three coupled coil system. Following the same approach as for the two coil system, it is straightforward to find

$$X = \begin{bmatrix} L_1 & M_{12} & M_{13} \\ M_{23} & L_2 & M_{23} \\ M_{13} & M_{12} & L_3 \end{bmatrix} - \begin{bmatrix} L_1\lambda & 0 & 0 \\ 0 & L_2\lambda & 0 \\ 0 & 0 & L_3\lambda \end{bmatrix}. \tag{6.54}$$

The characteristic equation becomes a third-order polynomial, which is equal to

$$(1-\lambda)^3 - (1-\lambda)(k_{23}^2 + k_{12}^2 + k_{13}^2) + 2k_{12}k_{23}k_{13} = 0, \tag{6.55}$$

where $M_{ij} = k_{ij}\sqrt{L_i L_j}$ was used. Since the coupling factors are $k_{ij} < 1$, the rightmost term of (6.55) being a product of three coupling factors maybe considered approximately zero, which allows one to easily obtain an approximate solution for the natural modes of the system. In this case, one can see that $\lambda = 1 \Rightarrow \omega = \omega_o$ is one solution of (6.55) and therefore in the system of three coils the frequency of one of the natural modes is approximately independent of the coupling factors of the coils. This solution is exact in the case where at least one pair of the coils is uncoupled, i.e., when at least one of k_{12}, k_{23}, or k_{13} is zero. This is for example the case where two of the coils are placed far from each other and the third one is used as a relay placed in a position between the other two coils. This is an interesting result because the frequency of this mode remains constant as the distance between the coils changes, which makes the design of such a system very attractive. It is left as an exercise to the reader to find a solution for the three natural modes as a perturbation of the solution corresponding to the one obtained by ignoring the last term of the characteristic equation.

The use of relay resonators placed in intermediate locations between the transmit and receive resonators provides a natural way to increase the transmission range of the system. In [170], for example, a vision of a multiresonator space where power can be coupled to a multitude of devices exploring one or more relay resonators placed around the walls of a space is presented, shown in Figure 6.17. Metamaterial arrays of resonators may also be used as lenses in order to relay power between two resonators with increased efficiency [171]. An experimental

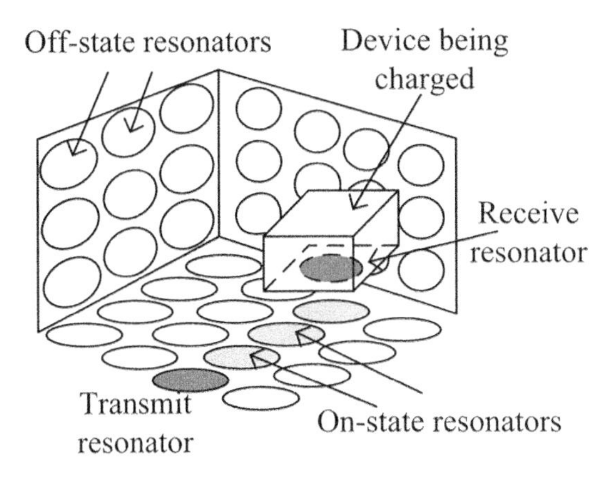

Figure 6.17 System concept of any-hop wireless power transmission.

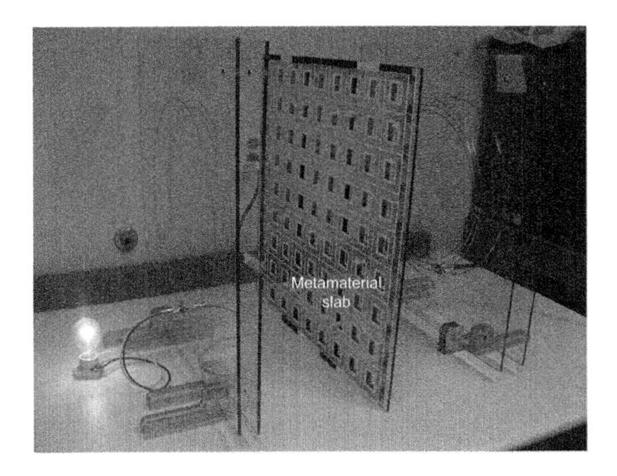

Figure 6.18 Metamaterial slab used to increase wireless power transmission efficiency of a resonant inductive coupling system demonstrator used to power a 40 W light bulb [171]. Photo courtesy of Dr. Bingnan Wang, Mitsubishi Electric Research Laboratories (MERL)

demonstrzation of a metamaterial slab used to focus the transmitted power of a resonant inductive wireless power transfer system is shown in Figure 6.18. The presence of the slab focuses the power toward the receiver, which is demonstrated by a larger intensity in the lightbulb.

Furthermore, the efficiency can be optimized by employing phased array concepts with multiple transmitters and receivers. The mathematical description of a multiple transmitter system optimization was formulated as a convex optimization problem in [156]. Convex optimization principles allow one to solve the underlying problem in an efficient and fast manner and facilitate the implementation of optimization algorithms in practical and commercial systems.

It is common to use four-loop structures such as the one shown in Figure 6.19a where two parasitic transmitting and receiving coil loops are used between the

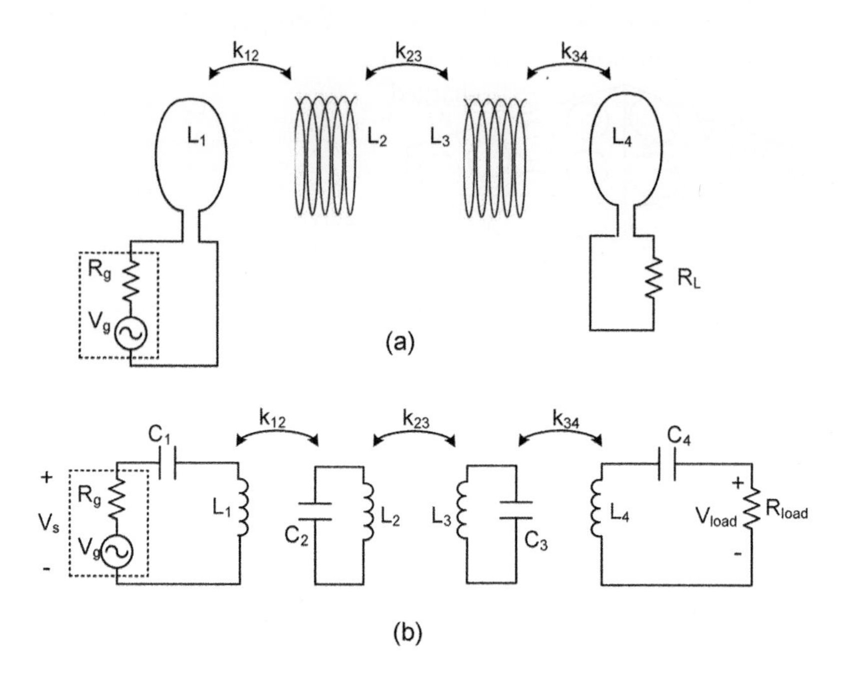

Figure 6.19 Resonant inductive coupling system with excitation loop and a load loop: (a) general scheme and (b) circuit model.

source and load loops [154]. This system has four natural modes that can be calculated following the previously descrbied methodology [169]. The two middle coils are designed to resonate at the same frequency as the transmitting and receiving coils, and they allow flexibility in implementing impedance matching for both the transmit and receive coils and maximizing the power transfer efficiency. For a fixed set of parameters of the resonant inductive coupling system, it is possible to select a coupling coefficient k_{12} between the source loop and the transmitting coil and a coupling coefficient k_{34} between the receiving coil and the load loop that allows one to optimize the input and output impedance matching conditions. The equivalent circuit model for the system in Figure 6.19a assuming resonant inductive coupling is shown in Figure 6.19b.

We have seen that as the coupling increases, which can happen for example if the distance between the coupled coils is reduced, the different modes separate in frequency [154]. As a result, an originally tuned system for a certain value of coupling may operate at a significantly reduced efficiency once the coupling changes. Alternately, an application scenario where one or both coils are moving significantly may present a large variation in operating efficiency. In order to maintain a high efficiency, one may retune the transmit or receive resonators in order to shift the operating mode frequency to a desired value that corresponds to a high efficiency. This unavoidably leads to an increased complexity in the wireless power transfer system because one needs to introduce a sensing mechanism

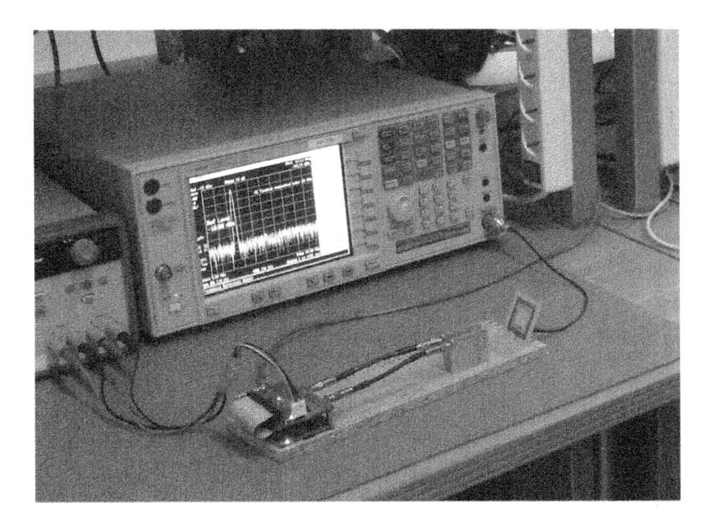

Figure 6.20 Software-defined configurable resonant inductive wireless power transfer platform based on a particle swarm optimizer (PSO) [174].

that tracks the operating efficiency, a feedback or feed forward link that can provide to the transmitter or receiver information about the efficiency and a tuning resonator functionality and control circuitry in order to implement the desired frequency tuning. Such tuning as well-adaptive tuning of wireless power transfer efficiency has received significant attention in the literature, for example in [172, 173]. In [174], for example, a software-defined radio (SDR) platform based on a Raspberry B+ evaluation module was used to implement a tunable 13.56 MHz resonant inductive wireless power transfer system, as shown in Figure 6.20. A tunable capacitor bank was controlled by the SDR platform in order to change the resonance frequency of the system based on the results of a particle swarm optimization (PSO), which was implemented offline in a Matlab environment.

6.4 Capacitive Power Transfer

An alternative way of using the magnetic field in order to transfer power wirelessly is to use the electric field. This is known as capacitive power transfer. A typical block diagram of a capacitive power transfer system is shown in Figure 6.21 [175]. In fact, Figure 6.21 shows a circuit topology of a resonant capacitively coupled system where the reactance of the coupled capacitors C_1 and C_2 has been compensated by the inductors L_1 and L_2. Similarly to the inductive coupled systems, there exist four different topologies corresponding to a parallel or series connection of the compensating inductors at the primary and secondary circuit. In practice, in addition to the two main capacitances there exist cross-coupling parasitic capacitances C_3 and C_4, as shown in Figure 6.21. One may transform the complicated coupled capacitor circuit with the parasitic capacitances into a

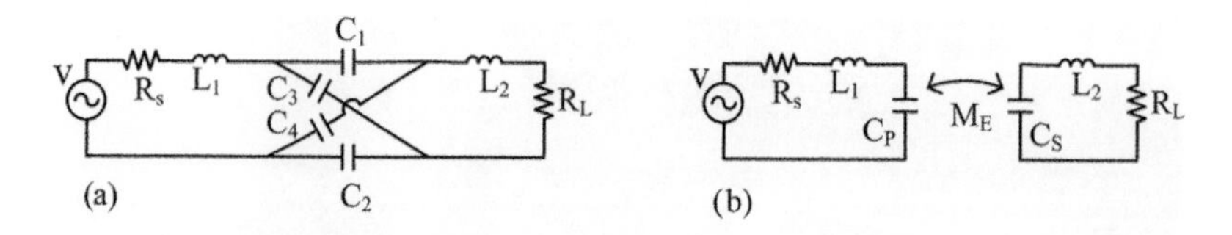

Figure 6.21 Resonant capacitive coupled wireless power transfer system: (a) circuit *schematic and (b) equivalent circuit.

circuit comprising two coupled capacitances C_P and C_S and at the same time define a capacitive mutual coupling coefficient M_E in a complete analogy with the inductive coupled systems. The resulting circuit capacitances and coupling coefficient are defined as follows [175]:

$$C_P = \frac{(C_1 + C_3)(C_4 + C_2)}{C_1 + C_3 + C_4 + C_2} \tag{6.56}$$

$$C_S = \frac{(C_1 + C_4)(C_3 + C_2)}{C_1 + C_3 + C_4 + C_2} \tag{6.57}$$

$$M_E = \frac{-C_3C_4 + C_1C_2}{C_1 + C_3 + C_4 + C_2} \tag{6.58}$$

and

$$k_E = \frac{M_E}{\sqrt{C_P C_S}}. \tag{6.59}$$

In this case, applying Kirchhoff's current laws in the system of Figure 6.21 one has

$$\begin{bmatrix} I_g \\ 0 \end{bmatrix} - \begin{bmatrix} G_g & 0 \\ 0 & G_L \end{bmatrix} \begin{bmatrix} V_1 \\ V_2 \end{bmatrix} = [Y] \cdot \begin{bmatrix} V_1 \\ V_2 \end{bmatrix}, \tag{6.60}$$

where

$$Y = \begin{bmatrix} G_1 + j\omega C_P \left(1 - \frac{\omega_o^2}{\omega^2}\right) & j\omega M_E \\ j\omega M_E & G_2 + j\omega C_S \left(1 - \frac{\omega_o^2}{\omega^2}\right) \end{bmatrix}. \tag{6.61}$$

The resonant capacitively coupled system therefore takes a dual form to the resonant inductive system of Figure 6.8.

Capacitive wireless power transfer systems were initially perceived for low power levels and at short distances in the order of 1 mm [176]. However, it was determined that both types of systems can achieve comparable efficiencies higher than 90% at kilowatt power levels, and generally there exist no guidelines to determine which type of system is more suitable for a certain power level, gap distance, and cost [176]. However, inductive power transfer systems are difficult to implement in application scenarios where power needs to be transferred through metal barriers due to eddy current losses and requires special shields to prevent electromagnetic interference (EMI) and magnetic cores to increase the coupling factors that correspondingly increase the cost. Capacitive coupled systems do not require magnetic cores, and because the electric field does not require a return path, it is easier to contain it between the capacitive plates and limit EMI [177].

6.5 Far-Field Wireless Power Transmission

When the distance between the transmit and receive antennas is large relative to the effective aperture area of the antennas, the wireless power transmission system operates in the far-field regime. We have seen in the introduction that in this case the Friis transmission formula holds and the efficiency of the radiating apertures η_{ap} is proportional to the square of the parameter τ, as shown in (6.7).

Following the work of Tesla, beginning in the 1960s, several far-field wireless power transmission systems have been developed, particularly in the United States and Japan. These systems comprised highly directive antennas transmitting a microwave beam on one side and large arrays of rectennas receiving the RF power and converting it to dc electrical power on the other side. There are several publications in the literature documenting these experiments, such as [121, 126, 178].

W. C. Brown demonstrated transmission of microwave power to a tethered helicopter in 1964 at Raytheon [126]. A photo of the helicopter carrying the receiving rectenna array is shown in Figure 6.22 and a more detailed photo of the "string" rectenna array is shown in Figure 6.23. The receiving array is an array of vertical strings of diodes separated by approximately a half wavelength, covering a 4 square foot area comprising 4480 IN82G diodes. It was capable of delivering a maximum dc output power of 270 W, which was sufficient to power the helicopter rotor.

In 1968, the solar power satellite concept was proposed by Glaser [126] . It consisted of capturing the sun's energy in a geosynchronous orbit and converting it into dc electrical power using solar panels and subsequently convert it into microwave power and transmit it to earth, where it can be converted back to dc electrical power. This activity has led to a milestone experiment of far-field wireless power transmission by Brown and Dickinson's team. They were able to demonstrate transmission of a microwave beam over 1 mile at 2.388 GHz at Goldstone, California, USA, in 1975 [179]. The Venus station 26 m diameter reflector antenna was used to transmit a microwave beam of up to 450 kW of power using a klystron generator. In the receiving side, a rectenna array was used comprising 17 subarrays of 270 dipole antennas each, placed above a ground plane at roughly a quarter wave distance and GaAs diode rectifiers. A photo of the experiment is shown in Figure 6.24. The efficiency of the radiating apertures of the system was approximately $\eta_{ap} = 11.3\%$. More than 30 kW of dc electrical power was received, corresponding to a RF–dc conversion efficiency of more than $\eta_{RFdc} > 80\%$ at the receiver.

Starting from the 1980s, many far-field wireless power transmission experiments were performed in Japan [178]. In 1992, a first trial employing a phased array transmitting antenna by Kyoto University and Kobe University,

Figure 6.22 Helicopter powered by a microwave beam flying 60 ft above the transmitting antenna. ©1984 IEEE. Reprinted with permission from [126]

demonstrated flying an airplane powered by a microwave beam at 2.411 GHz using a phased array with 96 GaAs amplifiers and 288 antennas grouped in three subarrays. A photo of the airplane and the transmit phased array are shown in Figure 6.25.

Following these milestone demonstrations, a plethora of far-field wireless power transmission experiments have been performed, exploring technological advances in power generation using, for example, magnetron and solid-state amplifier arrays, exploring antenna array concepts such as retrodirective arrays as well as technological advances in devices for rectifier circuits such as GaN technology [178].

6.6 RF-to-dc Conversion: the Rectifier

In the receiving side, the received RF signal from the antenna, coil, or capacitive radiator is converted back to a dc electrical signal. A nonlinear device is necessary in order to perform frequency translation from RF to dc. Typically a nonlinear resistive element is used, such as a diode or a transistor. These nonlinear devices

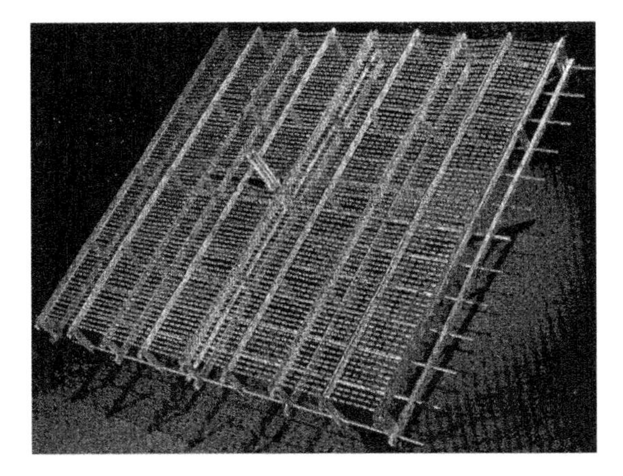

Figure 6.23 "String" rectenna used to power a helicopter in 1964. ©1984 IEEE. Reprinted with permission from [126]

Figure 6.24 Wireless power transmission experiment at Venus JPL cite. ©1984 IEEE. Reprinted with permission from [126]

are embedded in circuits called rectifiers that are optimized in order to maximize the RF–dc conversion efficiency η_{RFdc}. One should note, however, that in principle nonlinear reactive elements could also be suitable for rectification provided they generate a required mixing product from RF to dc; however, we are not aware of any such examples to date.

The most commonly used diode-based rectifier topologies are the series or shunt diode rectifier, but also voltage doubler circuits with two diodes or bridge rectifier circuits with four diodes. These topologies are shown in Figure 6.26. Voltage multipliers with multiple diode stages such as the voltage doubler topology are also used in order to perform rectification as well as maximize the dc output voltage.

Figure 6.25 Airplane powered by a microwave beam from a phased array antenna: (a) flight experiment, and (b) antenna array [178]. Photos courtesy of Prof. Naoki Shinohara, Kyoto University

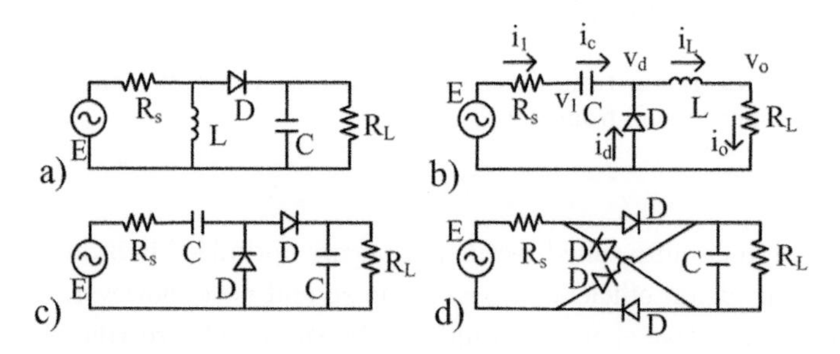

Figure 6.26 Diode rectifier topologies: (a) series, (b) shunt, (c) voltage doubler, and (d) bridge.

The nonlinear element acts as a switch, permitting current to flow through it only in one direction, and this functionality results in generating harmonics and dc power. In order to get a theoretical estimate of the maximum efficiency, one may proceed by considering an ideal switch with an "ON" state corresponding to a short circuit and an "OFF" state corresponding to an open circuit, taking care that the voltage and current values at the boundary of the two states must be the same. Then one applies Kirchhoff's laws and computes a time average of the voltage or current expressions over one period T of the RF signal. The results are used to derive an estimate of the RF–dc conversion efficiency. This methodology has been followed, for example, in [137, 180, 181]. More complicated models for the nonlinear device may be used by considering the nonlinear capacitance and series resistance of the diode [137] and nonideal open and short-circuit states for the switch [181]. Let us discuss the shunt diode rectifier following the approach by [180] considering an ideal switch model for the diode.

The available input power P_A, the dc output load power P_L, and the efficiency η_{RFdc} are

$$P_A = \frac{E^2}{8R_s} \tag{6.62}$$

$$P_L = \frac{V_o^2}{R_L} \tag{6.63}$$

$$n_{RFdc} = \frac{P_L}{P_A} = \frac{8R_s}{R_L}\left(\frac{V_o}{E}\right)^2. \tag{6.64}$$

We then proceed to write Kirchhoff's voltage law for the shunt rectifier circuit of Figure 6.26b:

$$v_1 + V_c = v_L + V_o = v_d, \tag{6.65}$$

where v_c and v_L is the voltage across the input capacitor C and output inductor L respectively. DC components are indicated by the capital V or I letters. It is assumed that the capacitor C has a sufficiently large value to present an RF short and that the inductor L has a sufficiently large value to present a dc short. Integrating over one period T, one obtains

$$\int_T v_1 dt + V_c T = \int_T v_L dt + V_o T \Rightarrow V_c = V_o. \tag{6.66}$$

The integral of the RF voltage across the inductor v_L vanishes since it presents a dc short. Furthermore, since no dc current flows through the resistor R_s (in the steady state) due to the dc blocking functionality of the capacitor C, the integral of v_1 also vanishes. Next we address the two states of the diode switch. Let us assume that the diode is in the "ON" state for a period $|t| < t_1$ and whereas the "OFF" state lasts from $t = t_1$ until $t = T - t_1$. In this case, during the "ON" state, one has

$$v_1(t) = -V_o \Rightarrow i_1(t)R_s = E\cos(\omega t) - V_o = E\cos(\omega t) - I_o R_L \tag{6.67}$$

and during the "OFF" state

$$i_1(t) = I_o \Rightarrow v_1(t) = E\cos(\omega t) - I_o R_s. \tag{6.68}$$

We enforce the condition that $v(t_1)$ must take the same value for the two expressions for the "ON" and "OFF" states,

$$-V_o = E\cos(\phi) - I_o R_s \Rightarrow E\cos(\phi) = I_o(R_s - R_L), \tag{6.69}$$

where $\phi = \omega t_1$ and $V_o = I_o R_L$. Next, taking the integral of v_1 over a period T, one has

$$\int_T v_1 dt = 0 \Rightarrow \int_{-t1}^{t1} v_1 dt + \int_{t_1}^{T-t_1} v_1 dt = 0. \tag{6.70}$$

Using (6.67) for the first integral of the left-hand side and (6.68) for the second, one obtains

$$-V_o 2t_1 + \frac{E}{\omega}\left[\sin(\omega T - \omega t_1) - \sin(\omega t_1)\right] - I_o R_s(T - 2t_1) = 0. \tag{6.71}$$

Rearranging and using $V_o = I_o R_L$, $\omega T = 2\pi$ and $\phi = \omega t_1$,

$$E\sin(\phi) = -I_o\left[\phi R_L + (\pi - \phi)R_s\right]. \tag{6.72}$$

Combining (6.69) and (6.72) one obtains

$$\tan(\phi) - \phi = \frac{\pi R_s}{R_L - R_s} \Rightarrow \frac{R_L}{R_s} = \frac{\sin(\phi) + (\pi - \phi)\cos(\phi)}{\sin(\phi) - \phi\cos(\phi)}. \tag{6.73}$$

Using the preceding result in (6.69), one can compute the output voltage V_o

$$V_o = I_o R_L = \frac{R_L E\cos(\phi)}{R_s - R_L} = \frac{E}{\pi}\left[\sin(\phi) + (\pi - \phi)\cos(\phi)\right] \tag{6.74}$$

and the efficiency

$$\eta_A = \frac{8}{\pi^2}\left[\ sin(\phi) - \phi\cos(\phi)\right]\left[\sin(\phi) + (\pi - \phi)\cos(\phi)\right]. \tag{6.75}$$

It is easy to very if by substitution that the maximum efficiency is obtained for $\phi = \pi/2$, resulting in

$$\eta_{Amax} = \frac{8}{\pi^2} \approx 81.1\% \tag{6.76}$$

with $R_L = R_s$. It is interesting to note that the same maximum efficiency $\eta_{Amax} = 8/\pi^2 \approx 81.1\%$ for $R_L = R_s$ is obtained for the series rectifier too [180]. We will see in the next sections that it is possible to obtain 100% by properly terminating the harmonics generated by the rectifier.

6.6.1 Time Reversal Duality

Amplifier and rectifier circuits have an inverse functionality of converting dc electrical power to RF and the opposite respectively. Strictly speaking, an amplifier requires both a dc and an RF input and therefore oscillator circuits should be considered instead of amplifier circuits. However, let us disregard this fact for the sake of simplicity and consider both oscillator and amplifier as dc-to-RF conversion devices. Consideration of these properties has led to the formulation time reversal duality principle [182]. According to the time reversal duality principle, any resonant amplifier can be transformed to a resonant rectifier of the same operating class, with the nonlinear switching device current and voltage waveforms of the rectifier being time-reversed versions of the corresponding waveforms of the amplifier. In mathematical terminology, let us consider a dynamical system such as an oscillator or an amplifier described by the N-dimensional state vector $\mathbf{x}$, $\mathbf{x} \in \mathbf{R}^{\mathrm{N}}$

$$\frac{d\mathbf{x}}{dt} = \mathbf{f}(\mathbf{x}(t), t), \tag{6.77}$$

where $\mathbf{f}$ is a nonlinear vector function. If one considers the reverse time variable $\tau = -t$, then

$$\frac{d\mathbf{x}}{d\tau} = -\mathbf{f}(\mathbf{x}(-\tau), -\tau). \tag{6.78}$$

One may consider (6.78) as a new dynamical system with state variables $\mathbf{y}$ in the forward time t [182] as

$$\frac{d\mathbf{y}}{dt} = \mathbf{g}(\mathbf{y}(t), t) \tag{6.79}$$

with $\mathbf{g} = -\mathbf{f}$ and $\mathbf{y}(t) = \mathbf{x}(-t)$. The equation $\mathbf{y}(t) = \mathbf{x}(-t)$ expresses precisely the fact that the state variables $\mathbf{y}$ of the second system are time-reversed versions of the state variables and $\mathbf{x}$ of the first system [182].

This has an important implication in wireless power transfer systems for the following reason. It is well known that there exist classes of switched power amplifier circuits that have a theoretical dc-to-RF efficiency of 100%, such as class-E, class-F, or class-F^{-1} [183]. The principle of operation of these classes is to design the voltage and current waveforms of the nonlinear switching device such that they are offset with each other so that their product, which corresponds to dissipated power, is equal to zero. The time reversal duality principle showed that one may design rectifier circuits based on the original amplifier circuits, having time-reversed voltage and current waveforms with respect to the original amplifier circuits and having theoretically 100% RF-to-dc conversion efficiencies.

Such rectifier circuits operating in GHz frequencies were successfully demonstrated in [181]. Specifically, a 2.14 GHz rectifier circuit based on a GaN HEMT class-F^{-1} amplifier had a power-added efficiency of 84% at an output power of 37.6 dBm. The rectifier had a measured RF–dc conversion efficiency of 85% for 10 W input power. A diode-based rectenna and active antenna oscillator circuit operating at microwave frequencies was reported in [184]. The rectenna was optimized to operate at 2.45 GHz with a high efficiency of 85% while as an oscillator the circuit was operating at 3.3 GHz with a low efficiency. This fact highlights the challenge in exploiting the time duality principle if one wants to design a circuit that operates both as a rectenna and as an active antenna oscillator. First, one has to tune both circuits to operate at the same frequency band, or in a more general concept at the desired operating bands as a rectenna

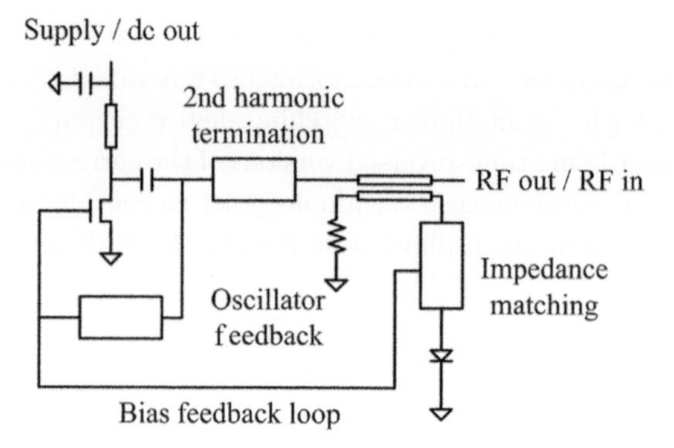

Figure 6.27 Schematic representation of a bidirectional rectifier and oscillator circuit operating at 2.45 GHz [185].

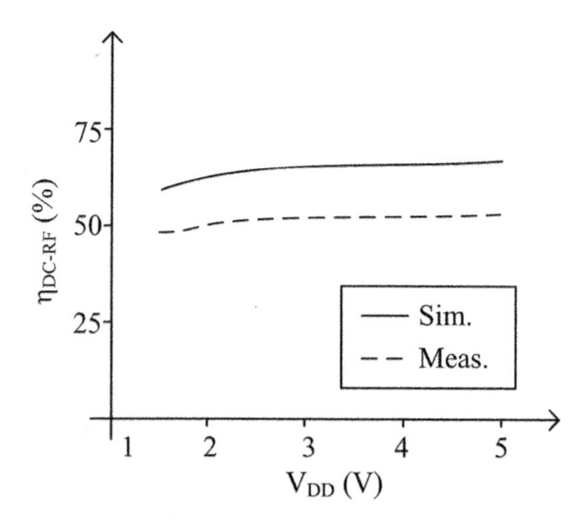

Figure 6.28 Performance of the oscillator mode of the circuit of Figure 6.27 [185].

and as an oscillator. Second, one has to tune both circuits to operate with high efficiency in both the rectenna mode and in the oscillator mode.

In [185], a transistor-based rectifier and oscillator circuit was designed operating at 2.45 GHz in both modes with an efficiency higher than 50%. A high-electron-mobility transistor (HEMT) device was used to implement the nonlinear circuits, and self-biasing of the gate of the device was explored in order to achieve a high efficiency at the oscillating mode. The circuit is shown in Figure 6.27 and the performance as an oscillator and as a rectifier is shown in Figures 6.28 and 6.29 respectively.

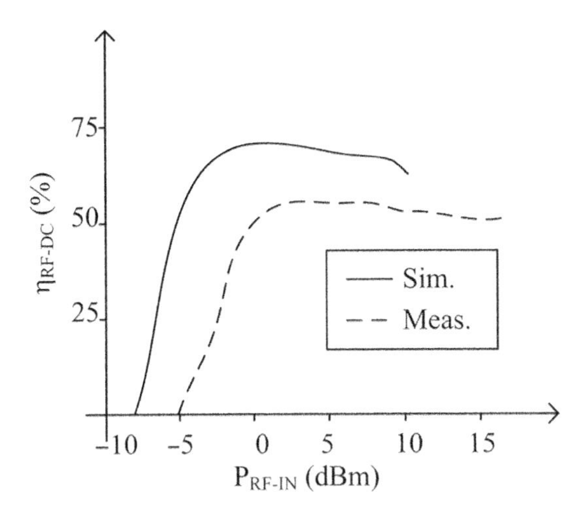

Figure 6.29 Performance of the rectifier mode of the circuit of Figure 6.27 [185].

6.7 Far-Field Wireless Power Transmission at Millimeter Wave Frequencies and Beyond

While the majority of far-field wireless power transmission circuits have been in the low GHz range due to the fact that it is easier and cheaper to generate high power and to obtain high-efficiency rectifiers at these lower microwave frequencies, operation at higher frequencies such as millimeter waves also has certain advantages.

Millimeter wave circuits allow for a small form factor due to the small wavelength, and therefore they permit the design of very compact circuits. More importantly, for the same reason they enable the implementation of a very large number of antenna elements, thus allowing one to design a very directive power transmitting antenna on one hand and a very directive receiving antenna or a large receiving surface of many individual subarray rectennas on the other hand. Therefore, it is possible to have a better control of the directive transmission of power and to minimize unwanted power transmission at undesired directions. These advantages may outweigh the high cost and less favorable performance of the available nonlinear devices in these frequencies.

Selected examples of millimeter wave far-field wireless power transmission circuits and systems include, for example, a 94 GHz rectenna array by JPL [186], and Class-F millimeter wave rectennas operating at 24 GHz and 60 GHz [142]. An example of a 24 GHz rectenna implemented in substrate integrated waveguide technology is shown in Figure 6.30 [187]. Its efficiency is shown in Figure 6.31, where one can verify that it is more difficult to obtain a high efficiency at comparable power levels at millimeter waves in comparison with rectennas operating at low GHz.

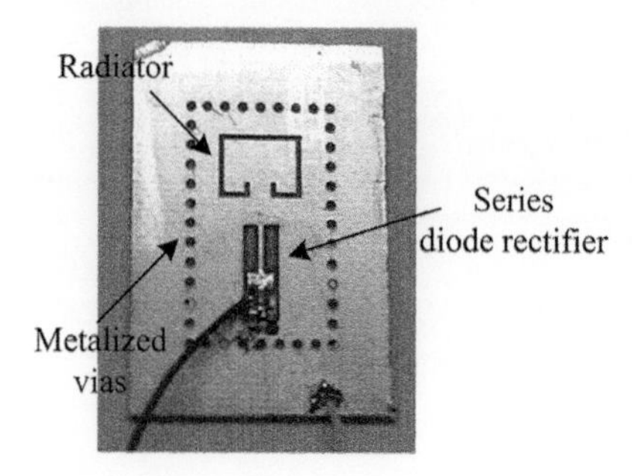

Figure 6.30 24 GHz SIW rectenna prototype. ©2013 IEEE. Reprinted with permission from [187]

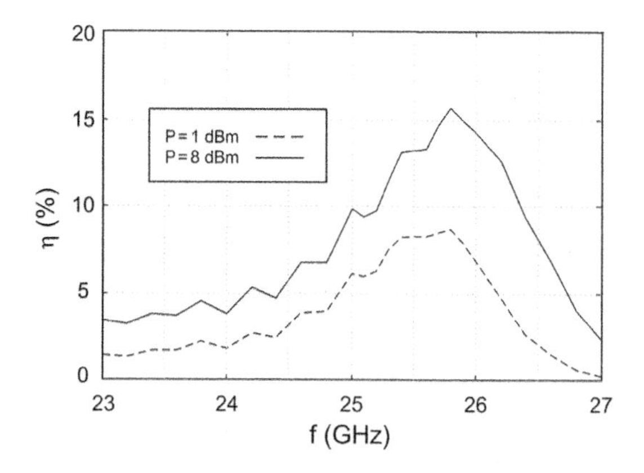

Figure 6.31 24 GHz SIW rectenna measured efficiency [187].

While high power generation is challenging at millimeter wave and toward THz frequencies, high power and highly directive optical sources are much easier to implement using lasers. Consequently, laser power transmission presents another exciting possibility for powering wirelessly devices. There have been successful demonstrations of transmitting power to a small aircraft using a laser, where solar cells have been used to receive the laser beam and convert it to electrical power [188].

6.8 Problems and Questions

1. What is the difference between inductive coupling and resonant inductive coupling?
2. Calculate the power transfer efficiency of a resonant inductive coupling system with $L_1 = L_2 = 5$ μH, $k = 0.3$ and $R_{load} = 100$ Ω operating at 13.56 MHz. Assume that the two coils have resistances $R_1 = R_2 = 2$ Ω.

3. Calculate the number of turns necessary to synthesize an inductance of 0.1 μH with a square printed coil with side length $D = 2$ cm using the modified Wheeler formula (assume $K_1 = 2.34$ and $K_2 = 2.75$).
4. Calculate the optimum value of the load resistance R_{load} for a nonresonant inductive coupling system with $L_1 = L_2 = 20$ μH and $k = 0.2$ at 13.56 MHz. Assume that the two coils have resistances $R_1 = R_2 = 2$ Ω.
5. Compute the maximum theoretical efficiency of a series diode rectifier with the circuit diagram shown in Figure 6.26.
6. Compute the maximum theoretical efficiency of a voltage doubler with the circuit diagram shown in Figure 6.26.
7. Compute an approximate solution for the eigenvalues of a three coil system with characteristic equation (6.55) as a perturbation of the solution obtained by setting $2k_{12}k_{23}k_{13} \approx 0$.

7 Electromagnetic Energy Harvesting

7.1 Introduction

The recent interest in batteryless sensors and sensors with increased energy autonomy has led to the concept of ambient electromagnetic energy harvesting or energy scavenging, where rectennas [127] are used as sources of dc electrical power by capturing available RF power from existing ambient low-power electromagnetic sources not intentionally transmitting to power a sensor and converting it to dc electrical power [132, 134]. We classified in Chapter 6 as energy harvesting the wireless power transmission scenarios operating in low power levels, which is typically the case when we are trying to power a sensor from an ambient RF signal.

7.2 Ambient Electromagnetic Energy

The average magnitude of the power density S of a plane electromagnetic wave with electric field E propagating in a medium of characteristic impedance η is

$$S = \frac{1}{2\eta}|E|^2. \tag{7.1}$$

Ambient electromagnetic power density S is typically measured in $\mu W/cm^2$. Alternatively, one may specify the electrical field strength $|E|$ in V/m. The characteristic impedance of air is equal to $\eta = 377\ \Omega$. The symbol η is typically used both for the characteristic impedance of a medium and for efficiency measures; however, the characteristic impedance is only used in this section and in section 7.4 in the book and, furthermore, the text should be clear enough to avoid any confusion between the two. It is straightforward to show that an electric field of 1 V/m in air corresponds to a power density of 0.26 $\mu W/cm^2$. The amount of ambient power density depends strongly on the application scenario and the environment that is under consideration. Therefore, it is bound to vary significantly between rural areas and cities where the number of operating wireless transmitters is very different. It addition, it is expected to strongly depend on frequency, based on the existing operating wireless communication systems, including TV, cellular, and Wi-Fi networks, to name a few. Measurement campaigns have appeared in the literature aiming to estimate the available power density in different scenarios [13, 14, 15]. As an example, [13] reported measured power densities in the 0.01–0.3 $\mu W/cm^2$ range for distances of 25 m–100 m from a GSM900 base station. Additionally, the authors of [13] report measurements of approximately one order of magnitude less than the GSM900 measurements, for an indoor measurement setup involving a Wi-Fi network. Such measurement campaigns are necessary as measurements depend strongly on the experiment setup and network traffic. In [14, 15, 189], the availability for harvesting digital TV signals was evaluated in Tokyo, Atlanta, and Seattle respectively. In [190], the amount of available energy from EM signals in the digital TV, the Global System for Mobile Communicatins (GSM) GSM, Wi-Fi, and 3G frequency bands is measured in the London area. In [191], a Worldwide Georeferenced Map containing the available EM signals levels was developed, where the recorded information may come from mobile devices or from laboratory equipment.

The available power that is collected by the radiating element of the rectenna is proportional to its effective aperture. The antenna effective aperture depends on the operating wavelength and gain of the antenna [120]. As an example, a square patch antenna with 3.4 cm side length and operating at 2.45 GHz has a simulated effective aperture of approximately 80 cm^2 [12]. There are several challenges in the design of the radiating element of a rectenna. One may consider various trade-offs between compact, dual polarized [12, 134] or circularly polarized rectennas [192] capable of receiving signals with arbitrary polarization and single-frequency band versus multiband [193, 194] and ultrawideband [132] designs aiming to simultaneously harvest electromagnetic power from as many as possible of the existing operating wireless systems.

The amount of dc power that can be harvested from the existing available power is proportional to the rectenna RF-to-dc conversion efficiency η_{RFdc}. The rectenna efficiency varies with the different rectifying circuit topologies and devices used, and it is dependent on the available power and the load resistance at the rectifier output. Low-power rectifier circuits typically use Schottky diodes to convert microwave power to dc power. Low and zero barrier diodes are required in order to rectify low-power input signals, similar to the ones used in detector applications. Nonetheless, rectifiers based on pseudomorphic high-electron-mobility transistor (p-HEMT) devices have also been proposed showing comparable efficiency values [185, 195]. Typically, envelope detectors and voltage multiplier or charge pump circuits based on diode or transistor devices are implemented, as shown in Figure 6.26. Reported rectenna efficiencies for available input power levels in the order of 10 μW (-20 dBm) are between 10% and 25%, and increase to 30% to 60% for available power levels of 100 μW (-10 dBm) [12, 132, 134, 193].

There have been recent studies in the literature aiming to estimate the available RF power density in ambient environments [13, 14, 190, 191, 196, 197].

Table 7.1 presents a nonexhaustive summary of the obtained results based mainly on the summary reported in [197] but also in [14, 191, 198].

Table 7.1 Selected reported ambient RF power densities [14, 191, 197, 198].

Reference	Year	Location	Power density nW/cm^2	DC power μW	Freq. band
Visser et al. [13]	2008	indoors	0.03	1,900	WLAN
Sample and Smith [189]	2009	Seattle, USA, outdoors		60	DTV
Olgun et al. [198]	2012	indoors	370	18	WLAN
Vyas et al. [15]	2013	Tokyo, JP, outdoors	> 26	0.93-29	DTV
Mimis et al. [199]	2015	Bristol, UK, outdoors	1,400	n/a	GSM900
Piñuela et al. [190]	2013	London, UK, outdoors	84	7400	GSM1800
Guenda et al. [191]	2014				

Consequently, electromagnetic energy harvesters have been proposed in the literature in order to enable wireless powering of sensors and other devices [134, 192, 196, 200]. Needless to say, the imagination, or rather, the ingenuity of the designer plays an important role in visualizing application scenarios where one can take advantage of the availability of ambient RF power and harvest it toward powering of sensors. For example, Figures 7.1 and 7.2 show two laboratory experiments, performed in 2009, where we investigated whether we could capture some of the RF energy leaking from a microwave oven or when a mobile phone is receiving a call. We used a rectenna comprising an ultrawideband printed monopole antenna connected with RF connectors to a series diode rectifier. In both cases, we were able to measure an output dc voltage of about 200 mV over a 2 KΩ resistor corresponding to approximately 20 μW or harvested power at a distance of about 1 m from the microwave oven and the phone. One may appreciate the difficulty in drawing reliable conclusions from such experiments due to the large number of parameters that affect the experiment, such as, for example, the propagation scenario, the state of the microwave oven (i.e., whether there is something inside the oven), the RF transceiver inside the phone, the antenna and the rectifier used in the rectenna, the propagation environment, etc.

Table 7.2 lists some examples of electronic devices and the amount of power that is required to power them up [197].

7.3 Low-Power Rectifier Circuits

As we have seen in Section 6.6, there exists a significant amount of literature regarding the theoretical performance analysis of rectifier circuits. It is possible to classify rectifier detector circuits into small signal and large signal detectors depending on the operating conditions of the nonlinear devices [202, 203].

Figure 7.1 Harvesting the leaked signal from a commercial kitchen microwave.

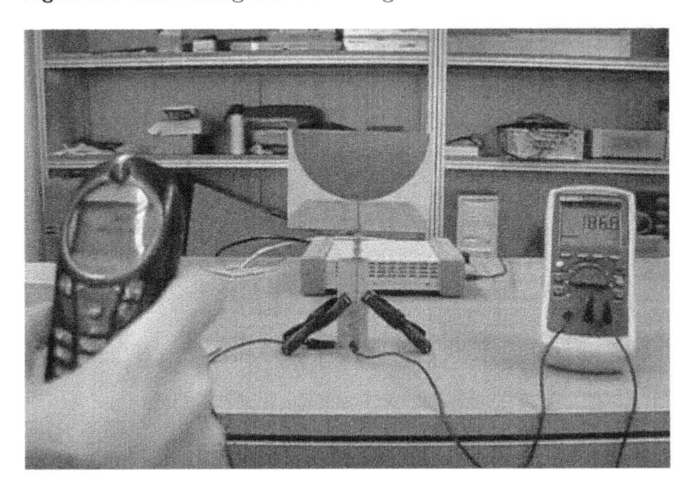

Figure 7.2 Harvesting the signal from a mobile phone.

In the latter case, the nonlinear device behaves as a switch, and we have seen in Section 6.6 that we need to model separately the different switching states. In [137, 141], a diode model including a series resistance and a parallel combination of a nonlinear resistance and capacitance was used to model a single shunt diode rectifier connected to a load. The voltage across the diode nonlinear resistance was modeled as a constant voltage drop when the diode is on and a harmonic signal containing a dc term and a fundamental frequency term when the diode is off. A harmonic expansion using dc and fundamental frequency terms was used for the nonlinear capacitor, the linear diode resistance, and the total voltage across the shunt diode and rectifier load. The analysis for a harmonically terminated diode and transistor rectifiers was presented in [181], whereas in [180]

Table 7.2 Selected electronic devices and their operating power [197].

Device	dc power (μW)
Small calculator	2
Monza 6R RFID tag [201]	6.2
Wristwatch	20
Digital thermometer	20
Smoke detector	55
Computer mouse	20,000
LED	60,000
Smartphone	500,000

fundamental limits of different diode-based rectifier circuit RF-dc conversion efficiency were derived.

In the former case, it is possible to apply a series approximation in the current-to-voltage characteristic of the nonlinear device in order to obtain meaningful and intuitive results about the rectifier circuit behavior [202]. A harmonic expansion using modified Bessel functions is often used for the diode current [204, 205, 206], while the dc power is evaluated by averaging over the input signal period [204, 205, 206, 207].

Let us focus on the simplest rectifier circuit, the series diode rectifier, shown in Figure 7.3. In the case of a small voltage signal applied across the diode terminals as shown in Figure 7.4a, the rectifier operation depends on the slope and the curvature of the current-to-voltage characteristic of the diode at the dc bias point [203]. The dc voltage output of the circuit is proportional to the power of the input signal, and the circuit is known as a "square law" detector. One characteristic measure of diode detectors is the current responsivity β

$$\beta = \frac{1}{2}\frac{d^2i/dv^2}{di/dv} \tag{7.2}$$

given by the half of the ratio between the second and the first derivatives of the current-to-voltage characteristic at the dc operating point. The current responsivity is equal to the dc current that is generated by the detector over the average RF power that is absorbed by the detector when the nonlinear current-to-voltage characteristic of the detector is approximated by a power series where only terms up to second order are considered [208]. The dc output power of the detector and the detector RF-to-dc conversion efficiency are thus proportional to β^2. As we will see in the next paragraphs, another important characteristic is the input resistance of the detector, which is critical in order to be able to implement a good impedance matching circuit between the rectifier and the antenna. Finally, the asymmetry of the current-to-voltage characteristic is important in order to be able to operate the rectifier at zero dc bias. In other words, a high responsivity at zero dc bias voltage and an input resistance around 50 ohms are the characteristics of an ideal rectifier circuit.

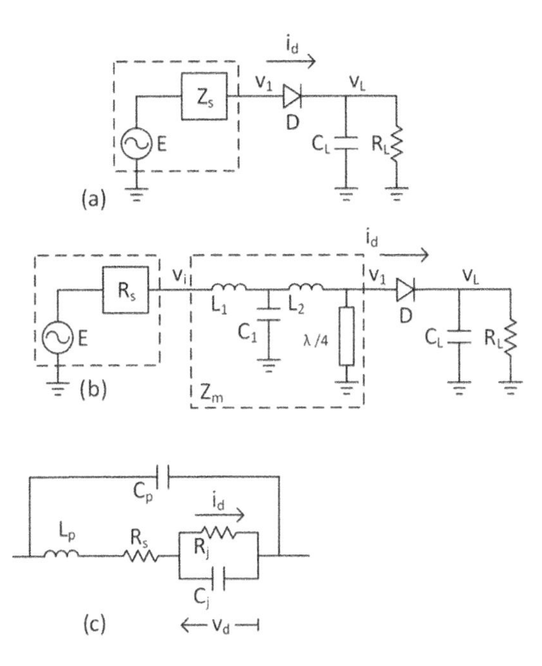

Figure 7.3 Block diagram of the rectifier setup in harmonic balance: (a) general model of the source, rectifier, and output filter; (b) model including source, matching network with harmonic termination, rectifier, and output filter; and (c) Schottky diode model. ©2016 IEEE. Reprinted with permission from [209]

In the case of a large voltage signal applied across the diode terminals as shown in Figure 7.4b, the diode conducts current only at a fraction of the input signal period and the current follows the peaks of the applied signal, resulting in a linear relationship instead between the output voltage and the input voltage to the rectifier [203].

Let us look at the series diode rectifier of Figure 7.3 in more detail. In a preliminary setup, a source with a desired harmonic impedance profile Z_s is connected in series with a Schottky diode D, which is followed by a shunt capacitor C_L and resistive load R_L, shown in Figure 7.3a. The rectifier efficiency η_A is defined as the ratio of the dc power P_L delivered to the output load R_L over the average available RF power from the source P_A.

$$\eta_A = \frac{P_L}{P_A}. \tag{7.3}$$

As we have seen in (6.23), alternatively an efficiency expression η using the input RF power can be used.

A harmonic balance simulation with seven harmonics has been set up in a commercial simulator. A nonlinear Schottky diode model has been considered corresponding to the Skyworks SMS7630-040 diode, which was selected for the analysis. The main parameters of the model are a nonlinear resistance R_j, a nonlinear capacitance C_j, and a series resistance R_s, shown in Figure 7.3c. The diode

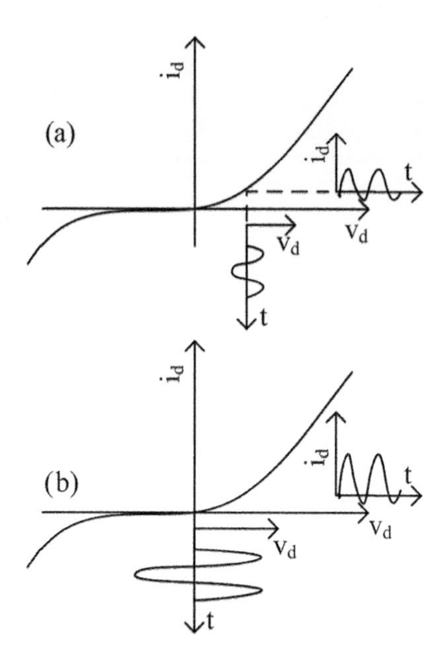

Figure 7.4 Diode rectifier/detector operation: (a) small signal and (b) large signal [202].

saturation current is I_s. Additionally, the diode has a breakdown voltage V_B. The diode package parasitics consist of an inductor L_p and a capacitor C_p.

Let us consider an ideal diode with $R_s = C_j = 0$, $V_B = 100V$, and $L_p = C_p = 0$. Furthermore, the load capacitance is sufficiently large ($C_L = 10$ nF) so that V_L consists only of dc voltage. In a first simulation, the source impedance is set to a real value $Z_s = R_s = 10$ KΩ that is constant at all frequencies, i.e., dc, the fundamental and the harmonics of the fundamental. The value of Z_s was selected arbitrarily and does not have effect in the qualitative behavior of the obtained results. The load resistance $R_L = rR_s$ is optimized to maximize the RF-dc conversion efficiency for an input signal with available power -20 dBm. In Figure 7.5, the efficiency n_A is plot versus the available input power P_A. It is seen that given enough input RF power, a maximum efficiency limit of 48.5% is obtained. If one sets the dc source impedance to 0Ω and performs the same experiment, one obtains a maximum efficiency of 80.9% for a sufficiently large input power. This is consistent with the theoretical limit obtained in [180]. Furthermore, proper harmonic termination can lead to a maximum theoretical efficiency of 100% as noted by [180] and demonstrated by [181]. Setting the source impedance to zero or to infinity at harmonic frequencies results in a maximum simulated efficiency of 99.5%, as shown in Figure 7.5. It should be emphasized that the preceding maximum efficiency values require an input signal with sufficiently large available average input RF power, and they are reduced by the diode series resistance, nonlinear capacitance and breakdown voltage, and the diode package parasitics, as well as losses in the source impedance network.

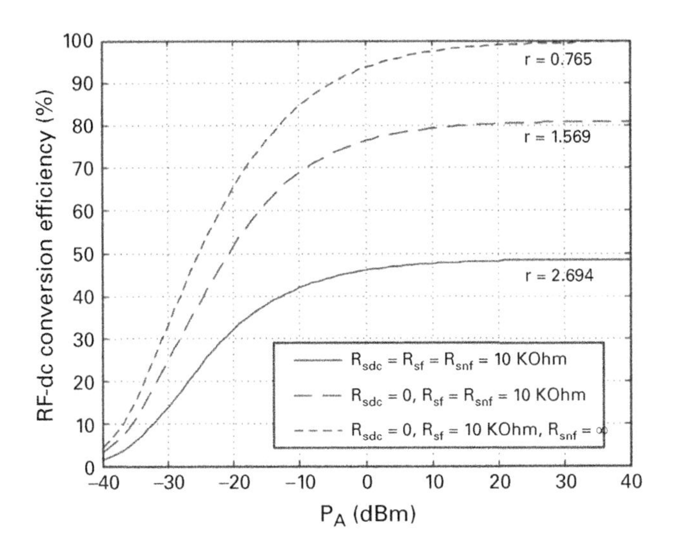

Figure 7.5 Simulated RF-dc conversion efficiency versus available input power of a continuous wave (CW) input signal.

The RF-dc conversion efficiency depends on the load resistance R_L. Figure 7.6 shows the obtained efficiency values versus R_L for an input available power of -20 dBm of a continuous wave (CW) signal for the three cases of different source impedance terminations considered in Figure 7.5. The peak efficiency values of each of the curves of Figure 7.6, correspond to the efficiency value listed in Figure 7.6 for $P_A = -20$ dBm. It can be seen that the optimal load leading to the peak efficiency is strongly affected by the source impedance at dc and the harmonics of the fundamental frequency.

Let us focus on the effect of the input signal on the optimal load and RF-dc conversion efficiency. In order to derive an approximate expression for the rectifier efficiency under small signal excitation, it is assumed

$$v_1(t) = V_{10} + \sum_{n=1}^{N} V_{1n} \cos(n\omega_o t) \approx V_{11} \cos(\omega_o t) \tag{7.4}$$

$$v_L(t) = V_{L0} + \sum_{n=1}^{N} V_{Ln} \cos(n\omega_o t) \approx V_{L0} \tag{7.5}$$

and

$$i_d(t) = I_{d0} + \sum_{n=1}^{N} I_{dn} \cos(n\omega_o t) \approx I_{d0} + I_{d1} \cos(\omega_o t). \tag{7.6}$$

The first equation (7.4) relies on the fact that the stub of the matching network (Figure 7.3b) leads to zero dc and even harmonic voltage components at the input of the diode, and it is assumed that odd harmonic voltage components are very small compared to the fundamental voltage. Similarly, it is assumed that

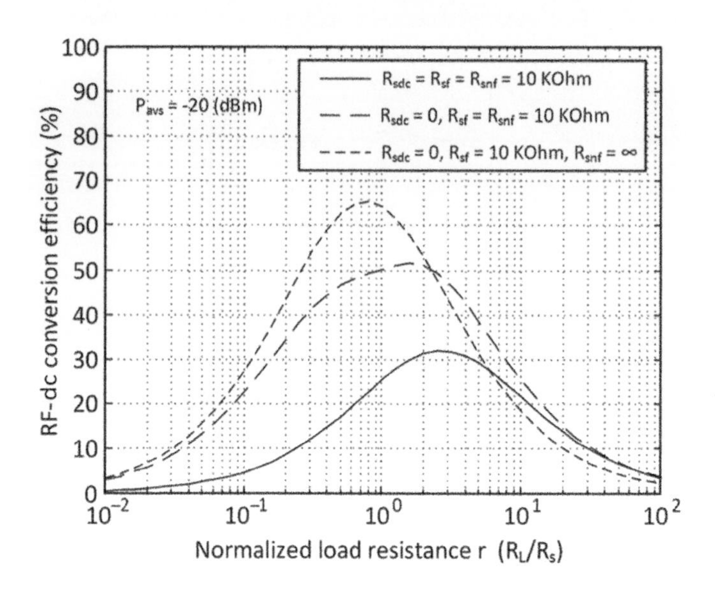

Figure 7.6 Simulated RF-dc conversion efficiency versus available input power of a CW input signal.

the output capacitor C_L minimizes the fundamental and harmonic components of the output voltage v_L. The phasors of the harmonic expressions are considered constant with time as a CW excitation has been considered.

Assuming $R_s = C_j = 0, V_B = 100V$, and $L_p = C_p = 0$, the diode current is

$$i_d(t) = I_s \left[e^{\alpha(v_1 - v_L)} - 1 \right], \tag{7.7}$$

where $\alpha = 1/nV_T$, $V_T = kT/q$ is the thermal voltage, n the diode ideality factor, k the Boltzman constant, T the junction temperature, and q the electron charge. It is easily verified that α is proportional to the current responsivity of the diode $\alpha = 2\beta$. Using (7.4) and (7.5) and the modified Bessel function of the first kind B_n series expansion [205]

$$e^{z\cos(x)} = B_o(z) + 2\sum_{n=1}^{+\infty} B_n(z)\cos(nx), \tag{7.8}$$

one obtains for the rectifier diode dc current

$$I_{d0} = I_s \left[e^{-\alpha V_{LO}} B_o(\alpha V_{11}) - 1 \right] \tag{7.9}$$

and for the rectifier diode RF current at the fundamental frequency ω_o

$$I_{d1} = 2I_s e^{-\alpha V_{LO}} B_1(\alpha V_{11}). \tag{7.10}$$

The input RF impedance of the rectifier is then equal to

$$R_1 = \frac{V_{11}}{2I_s e^{-\alpha V_{LO}} B_1(\alpha V_{11})} \Rightarrow x_1 = \frac{z}{2e^{-y}B_1}, \tag{7.11}$$

where we normalized the various parameters as $y = \alpha V_{L0}$, $x_1 = \alpha I_s R_1$, and $B_1 = B_1(z)$ with $z = \alpha V_{11}$. Using $V_{L0} = I_{d0} R_L$, one obtains the dc output voltage from (7.9)

$$V_{L0} = I_s R_L \left[e^{-\alpha v_L} B_o(\alpha V_{11}) - 1 \right] \Rightarrow y = x \left[e^{-y} B - 1 \right], \tag{7.12}$$

where again we normalized $x = \alpha I_s R_L$ and $B = B_o(z)$. It is possible to solve (7.12) for the output dc load voltage y using the Lambert W function [206]

$$y = W_o(x e^x B) - x. \tag{7.13}$$

Due to the even symmetry of the modified Bessel function of the first kind of order 0, $B = B_o(\alpha V_{11}) = B_o(\alpha |V_{11}|) > 0$, the argument of the Lambert function in (7.13), is always positive and the principal branch W_o of the Lambert function is used. The output dc power is $P_{L,dc} = V_{L0}^2 / R_L$, and using (7.13) it is calculated as

$$P_{L,dc} = \frac{I_s}{\alpha} y(B e^{-y} - 1). \tag{7.14}$$

The value of $y = y_m$, which leads to a maximum dc output power, is calculated by taking the derivative of (7.14) with respect to y and setting it equal to zero. It is straightforward to find that y_m fulfills

$$B e^{-y_m}(1 - y_m) = 1 \Rightarrow e^{1-y_m}(1 - y_m) = e B^{-1}. \tag{7.15}$$

Using the definition of Lambert function $q = p e^p \Rightarrow p = W(q)$, one has

$$y_m = 1 - W_o(e B^{-1}), \tag{7.16}$$

where again the principal branch of the Lambert function is used due to the fact that $B > 0$. The optimum value of the load $x = x_m$ is obtained from (7.15) and solving (7.12) for x as $x_m = 1 - y_m$, or

$$x_m = W_o(e B^{-1}). \tag{7.17}$$

The input impedance x_{1m} corresponding to the optimum load x_m becomes

$$x_{1m} = \frac{z_1 B W_o(e B^{-1})}{2 B_1}. \tag{7.18}$$

If $V_{11} \to 0$, then $B \to 1$ and $x_m \to W_o(e) = 1$. Furthermore, $B_1 \approx z/2$, resulting in $x_{1m} \to 1$ as well. In the event that $V_{11} \to +\infty$, then $e B^{-1} \to 0$ and $x_m \to 0$. This demonstrates that at the low input power limit, the optimum load of the rectifier circuit depends on the diode saturation current $x_m \to 1 \Rightarrow R_{Lm} \to (\alpha I_s)^{-1}$ and is reduced with increasing average input signal power. Due to the monotonic nature of W_o, one has $0 \leq x_m \leq 1$. In (7.17), we have derived a closed-form expression for the load value corresponding to maximum efficiency as

$$R_{Lm} = \frac{W_o(e B_o^{-1}(\alpha V_{11}))}{\alpha I_s}. \tag{7.19}$$

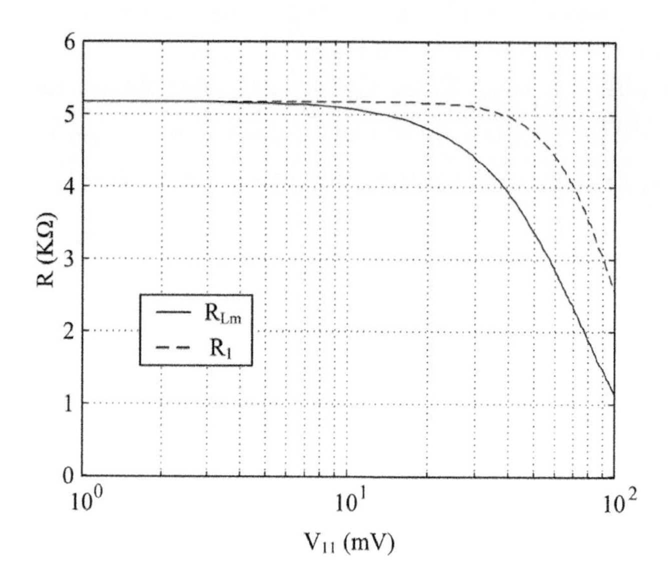

Figure 7.7 Calculated rectifier optimum load R_{Lm} and RF input resistance R_1 corresponding to the optimum load ($I_s = 5\ \mu$A).

The maximum dc output power becomes

$$P_{Lm} = \frac{I_s}{\alpha}\frac{y_m^2}{1-y_m} = \frac{I_s}{\alpha}\frac{(1-x_m)^2}{x_m} = \frac{I_s}{\alpha}\frac{\left[1-W_o(eB_o^{-1}(\alpha V_{11}))\right]^2}{W_o(eB_o^{-1}(\alpha V_{11}))}. \tag{7.20}$$

The dc output voltage is

$$V_{Lm} = \frac{1-W_o(eB_o^{-1}(\alpha V_{11}))}{\alpha}. \tag{7.21}$$

Finally, the input RF impedance of the rectifier at the optimum load becomes

$$R_1 = \frac{V_{11}B_o(\alpha V_{11})W_o(eB_o^{-1}(\alpha V_{11}))}{2I_sB_1(\alpha V_{11})}. \tag{7.22}$$

The calculated optimum load R_{Lm} and optimum input resistance R_1 versus the RF amplitude V_{11} of a CW signal using (7.19) and (7.22) is shown in Figure 7.7, for $I_s = 5\mu$A, which corresponds to the SMS7630 diode ($V_T = 25.85$ mV at 300 K). In this case, both the optimum load and the input impedance of the rectifier corresponding to the optimum load to the limit of small input power become $R_{Lm} = R_{1m} = 5.43\ k\Omega$. This high input resistance value poses a significant challenge to the designer trying to implement an impedance matching circuit over a wide bandwidth at low input power levels.

7.4 Nonlinear Optimization of Rectenna Circuits

Depending on the targeted application and attending to the maximum harvesting range and available frequency sources, different operation frequencies may be chosen for the rectenna element. The antenna and the rectifier elements have to

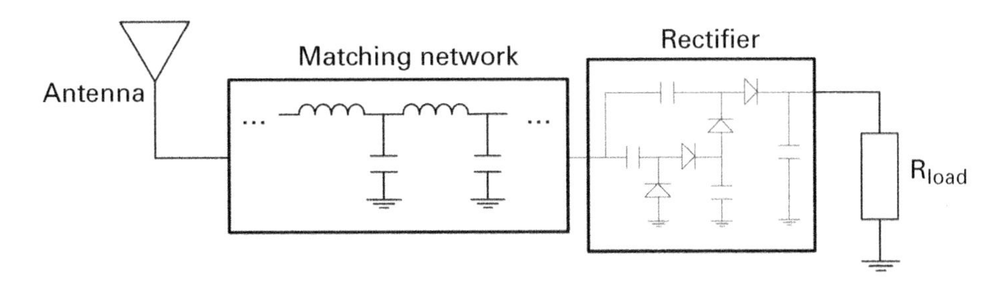

Figure 7.8 Schematic representation of the building blocks of a rectenna element.

be designed in order to maximize the amount of dc power that can be obtained at the selected operation frequency. The antenna has to be designed to operate and have good performance at the selected frequency, and the rectifier circuit has to be optimized to have its maximum RF-to-dc conversion efficiency at this frequency.

In the simplest scenario, one needs to design a single-frequency band rectenna. However, in some cases, it may be of interest to have dual-band, triple-band, and in general multiple-band or broadband rectenna designs in order to harvest energy from different frequency sources. The optimized design of a rectenna requires the simultaneous synthesis of several elements: (1) antenna, (2) rectifier circuit, and (3) matching network (Figure 7.8). In the case of the antenna element, it has to be designed to operate at the selected frequency bands leading to single-band, multiple-band, or broadband antenna designs. As previously mentioned, in order to maximize the RF-to-dc conversion efficiency of the rectenna, an adequately designed matching network has to be selected and optimized aiming to maximize the power transfer from the antenna port to the rectifier circuit. The optimum design of the matching network requires consideration of both the antenna and the rectifier circuit in the design stage. For a specific load R_L and for a selected topology of the rectifier circuit and of the antenna element, the matching network can be optimized to maximize the RF-to-dc conversion efficiency of the rectenna.

In order to be able to consider the antenna and the rectifier simultaneously in the design stage, the antenna must be introduced in the nonlinear circuit simulation. This is done by considering the antenna as a loaded scatterer and introducing the Thevenin or Norton equivalent circuit of the antenna in the receiving mode using the theory developed in [210]. This approach applied in rectenna circuits was implemented in [12, 193]. In Figure 7.9, for example, the Thevenin equivalent circuit is used, comprising an open-circuit voltage source V_{oc} in series with an impedance Z_A. The antenna impedance Z_A in the receiving mode is equal to the antenna impedance in the transmitting mode and can be computed using an electromagnetic simulator, for example, or by measuring the s-parameters using a network analyzer. The open-circuit voltage V_{oc} can be computed by considering reciprocity theory [211]. If one considers a plane wave with vector amplitude $\mathbf{E_o}$ arriving at the antenna terminals from a direction (θ, ϕ),

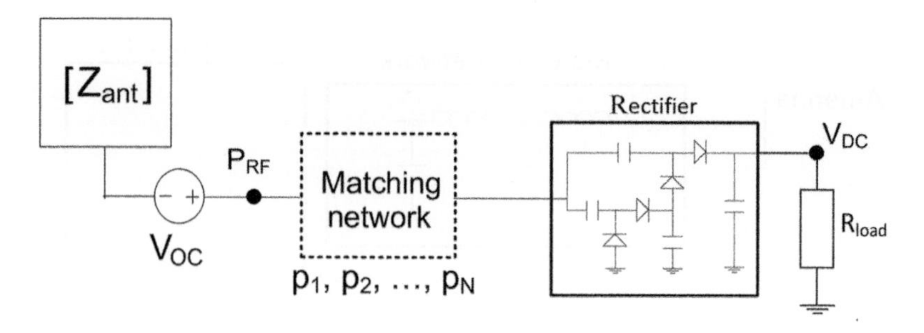

Figure 7.9 Rectenna simulation setup using the Thevenin equivalent representation for the receiving antenna.

then reciprocity theory dictates that the open-circuit voltage at the antenna terminals is given by

$$V_{oc}(\theta, \phi) = \frac{\lambda}{j60\pi} \mathbf{F}(\theta, \phi)\mathbf{E_o}, \tag{7.23}$$

where λ is the free space wavelength. $\mathbf{F}(\theta, \phi)$ is the electric far-field vectors of the antenna at the direction (θ, ϕ) when a unit current excitation is applied at its port [120, 210]. The antenna far-field can be also computed by an electromagnetic circuit simulator with the antenna operating in the transmitting mode. The impressed field has a magnitude $E_o = \sqrt{2\eta S}$, where S is the power density at the antenna terminals. This way, the Thevenin equivalent of the antenna can be introduced in a commercial circuit simulator in order to take into account the antenna structure when optimizing the rectifier. This setup allows one to separately address the antenna, which requires an electromagnetic analysis, and the rectifier, which requires a nonlinear circuit analysis, and subsequently link the two together.

Due to the nonlinear nature of the used rectifying devices, nonlinear simulation tools such time domain integration or harmonic balance must be employed for the analysis and optimization of the rectifier circuit. Harmonic balance provides an efficient simulation tool that can be combined with optimization goals in order to optimize a desired parameter, such as the maximizing the output dc voltage, or the RF to dc conversion efficiency of the rectifier.

In this last case, (7.3) can be used to define the RF-to-dc conversion efficiency η_a. It is noted that using the available power in the efficiency expression provides an equation that is easier to compute because the available power does not depend on the rectifier circuit or the matching network. In order to maximize η_a, a minimum value of RF-to-dc conversion efficiency at the desired frequency is imposed using optimization goals and the values of the matching network components $p_1, p_2, \ldots p_N$ as well as the rectifier load R_L is calculated in order to fullfil this goal.

The harmonic balance analysis of different rectifier circuits is shown in Figure 7.10. In order to compare different circuit topologies, a lossless matching

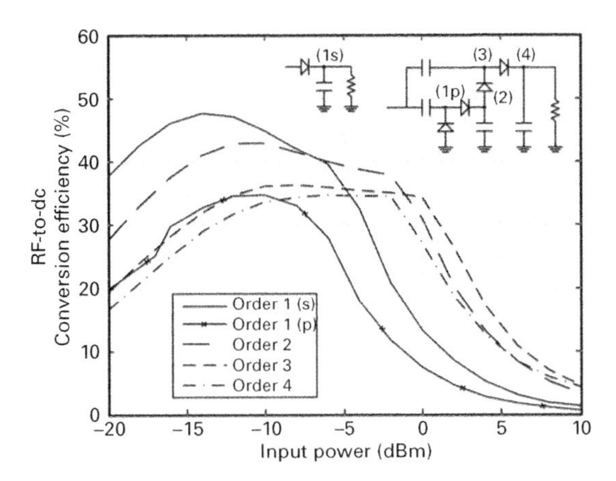

Figure 7.10 Rectifier efficiency versus the input available power for different rectifier circuits.

network was considered and the rectifier efficiency η_A was maximized for an available input power of −20 dBm. Five different topologies were considered, a single series diode rectifier, a shunt diode rectifier, and three voltage multiplier topologies with two, three, and four diodes respectively. The SMS7630 diode model was considered in the simulation. After the optimization at −20 dBm, the matching network parameters and the output load were fixed, the input power was sweeped from −20 dBm to 10 dBm. The obtained efficiency is plotted in Figure 7.10.

Based on the results, we can draw several conclusions regarding the performance of rectifier circuits. We can identify three regions based on the input power levels. First, at low input powers, the efficiency increases relatively linearly with the input power. Comparing the different circuit topologies, we can see that the single series diode rectifier has the higher efficiency, and as the number of diodes increases, the efficiency is reduced. This is intuitive because there is some power lost in each diode and therefore the more diodes one uses the more power is lost and the lower the efficiency becomes. However, as one increases the number of diodes, the output voltage of the rectifier increases. The designer is therefore presented with a trade-off between desired output load voltage and efficiency.

As the input power increases, the efficiency begins to reach a plateau. This is due partially to reflection losses associated with impedance mismatch between the rectifier and the matching network. Due to the nonlinear nature of the rectifier, the input impedance changes as the input power varies from its original value at −20 dBm where the matching network was designed.

Finally, as the input power increases further, there is a point where efficiency begins to drop suddenly. This is because the voltage across the diode begins to reach the diode breakdown voltage value, which results in a sudden increase in the current through the diode and significant power losses in the diode. Furthermore, the rectrifier circuits with fewer diodes reach this efficiency drop point faster than

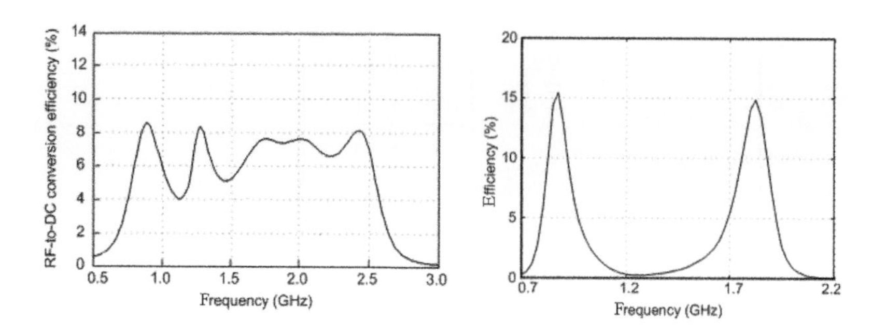

Figure 7.11 Multiband versus wideband rectifier efficiency.

the ones with a larger number of diodes. This is also intuitive because a larger voltage across the rectifier terminals is necessary to bring all the diodes to the breakdown voltage zone.

We have already seen in Figure 7.6 that there is an optimum load value that maximizes efficiency for a given input power level. Furthermore, the optimum load varies with the input power. We can see that due to the nonlinear nature of the rectifier circuits, the designer is faced with several challenges defined by the operating frequency, the operating power, and the output load. How to address these challenges is the focus of the next sections.

7.5 Multiband Rectifiers and Rectennas

We have seen that the available power density due to ambient RF sources is quite low. In order to be able to harvest a sufficient amount of power, one has to consider a large aggregate bandwidth covering either a number of different disjoint frequency bands or a large, wide bandwidth. The first scenario is suitable when the sources of RF power are known a priori, i.e., mobile phone bands, Wi-Fi bands, or TV bands. Consequently, one aims to design a multiband rectifier. The second scenario becomes attractive when the frequency of RF sources is not known a priori and one needs to design a rectifier with high efficiency over a wide bandwidth covering multiple frequency bands. The two cases are presented conceptually in Figure 7.11. The design process becomes essentially that of an impedance matching problem over a desired aggregate frequency bandwidth, with the additional constraints over the input power and output load impedance. Due to a fundamental theoretical limitation in the achievable impedance matching over a frequency band studied by Bode and Fano [212], which we will discuss in more detail in the next section, it is generally possible to achieve a higher maximum efficiency over a number of narrow frequency bands with a smaller aggregate bandwidth than over a wider continuous frequency band.

It is possible to compare different rectennas by expressing the obtained efficiency over either the input power density (in e.g. μW/cm^2) or the input power

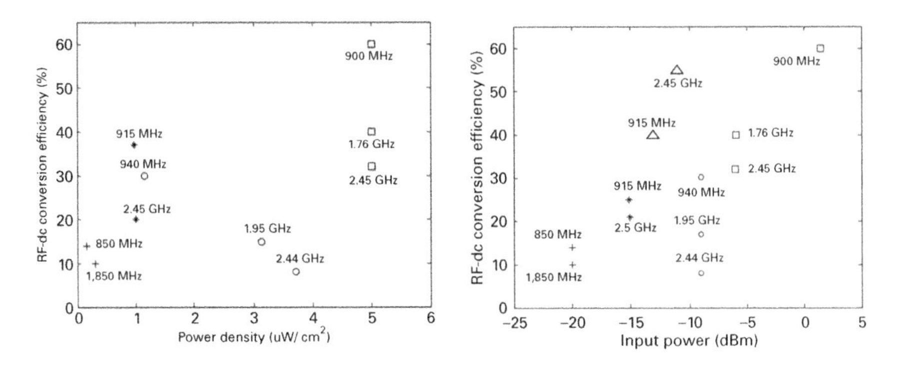

Figure 7.12 Selected published dual-band rectenna efficiencies plot versus (a) input power density (©2013 IEEE, reprinted with permission from [213]) and (b) available input power (○ [214], + [131], ∗ [213], □ [193], △ [215]).

(in e.g. dBm) at the rectifier antenna terminals. An example of such plots related to multiband rectennas is shown in Figure 7.12.

In the first case, when the input power density is considered one takes into account the antenna gain in the rectenna comparison and effectively considers both the antenna and the rectifier circuit combination, whereas in the second case the comparison is made only between rectifier circuits, therefore excluding the effect of the antenna circuit. Due to the fact that the antenna gain depends on the size of the antenna, one may want to additionally consider the size of the rectenna, when comparing different designs. In an attempt to define a figure of merit taking into account all these parameters, the authors introduced at the wireless power transmission student design competition of the 2011 IEEE Microwave Theory and Techniques Society (MTT-S), International Microwave Symposium (IMS), the following expression

$$\mathrm{FoM} = 10\log_{10}\left(\frac{P_L}{10}\right) - 10\log_{10}\left(\frac{D^2}{25}\right) \tag{7.24}$$

in dB of the rectenna dc output power P_L (μW) normalized over 10 μW, divided by the area calculated as the square of the largest dimension D (cm) of the rectenna normalized over 25 cm^2. The expression includes the size of the rectenna in an attempt to define an objective FoM figure. Although not widely accepted, it is indicative of the multiple challenges in designing a rectenna circuit.

7.6 Ultrawideband Rectifiers

The input impedance Z_e of a rectifier circuit contains a resistive component and a reactive, typically capacitive, component. Furthermore, both components vary with the input power. For a given input power level, it is possible to provide

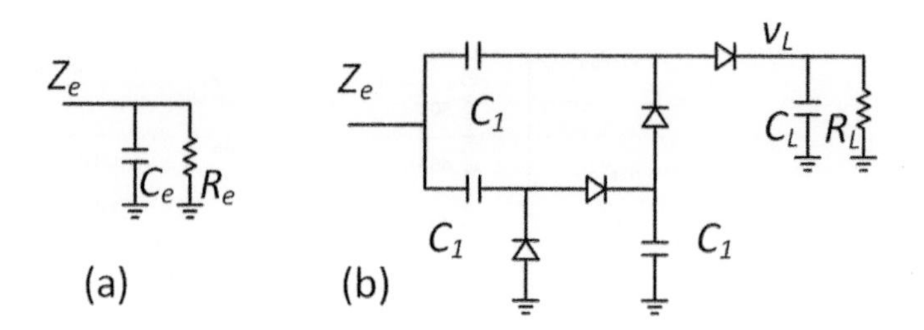

Figure 7.13 Equivalent circuit model of a rectifier: (a) shunt RC equivalent, and (b) charge pump rectifier. ©2017 IEEE. Reprinted with permission from [216]

the input impedance of a rectifier by defining an equivalent circuit model that comprises a shut resistor and a shunt capacitor, as shown in Figure 7.13.

The equivalent resistance and capacitance of rectifier circuits with a different number of diodes is simulated in Figure 7.14 [216]. The circuit schematic was the one shown in Figure 7.13b with a varying number of diodes $N = 1, 2, 4,$ and 6, where $N = 1$ corresponded to a series diode rectifier. The equivalent resistance and capacitance were extracted by minimizing the difference between the input impedance of the rectifier circuit and a shunt RC circuit over the frequency band 0.4–1.0 GHz. The Skyworks SMS7630 diode model was used, and capacitors $C_1 = 100$ pF and $C_L = 10$ nF were selected such that they presented an effective RF short within the frequency band 0.4–1.0 GHz. Finally, the load resistance was fixed at 1 KΩ.

The input resistance varies significantly with the input power whereas the input capacitance variation becomes more visible at higher input power levels. As we have verified in the theoretical analysis of the previous sections, the input resistance of the rectifier circuits is larger at low input power levels and for a smaller number of diodes. In fact, it appears that for the given diode model and circuit topology, a rectifier with six diodes already presents an input resistance close to 50 Ω at low input power levels of approximately −20 dBm, and therefore it should be easier to design an impedance matching network for this circuit. Although not visible from Figure 7.14, the input impedance depends strongly also on the load resistance.

Having obtained a simple equivalent circuit for the rectifier input impedance, we can address the problem of designing an impedance matching network over a desired frequency band. Precisely, the theory developed by Bode and extended by Fano [212] addresses this problem and derives that the minimum reflection coefficient magnitude $|\Gamma_m|$, which can be obtained when attempting to match a shunt RC load impedance comprising R_e and C_e, over a bandwidth B, using a lossless matching network, is limited by

$$|\Gamma_m| \geq e^{\frac{1}{2BR_eC_e}}. \tag{7.25}$$

Using the results of Figure 7.14 in (7.25), we can obtain an estimate of the minimum achievable theoretical reflection coefficient magnitude over a desired

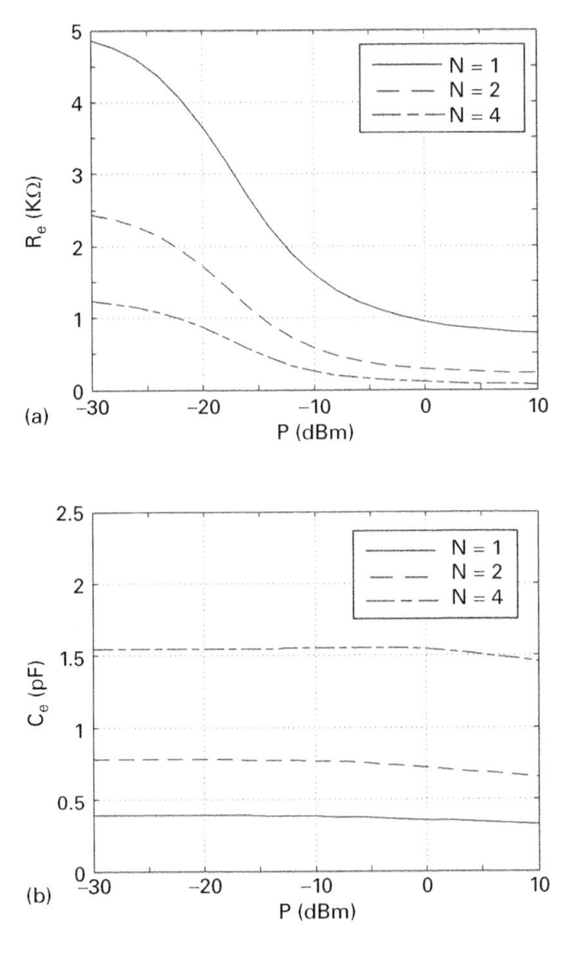

Figure 7.14 Equivalent input impedance of a rectifier circuit: (a) resistance and (b) capacitance [216].

bandwidth of 0.4–1.0 GHz. The result in shown in Figure 7.15, verifying that it is more challenging to design a wideband impedance matching network at low input power levels and for a rectifier with fewer diodes.

The theoretical results help select a rectifier topology for a given input power and output load resistance. The challenge remains, however, how to implement such an impedance matching network. There is a vast amount of literature related to filter design and impedance matching that the designer can apply. For example, a nonuniform transmission line is one topology that traditionally has been used for broadband impedance matching [217]. We demonstrated an octave band and a decade band rectifier based on a printed nonuniform microstrip transmission line using harmonic balance optimization [216]. First, a charge pump rectifier topology using four diodes was selected based on Figures 7.14 and 7.15. The design process consisted of considering a number of microstrip

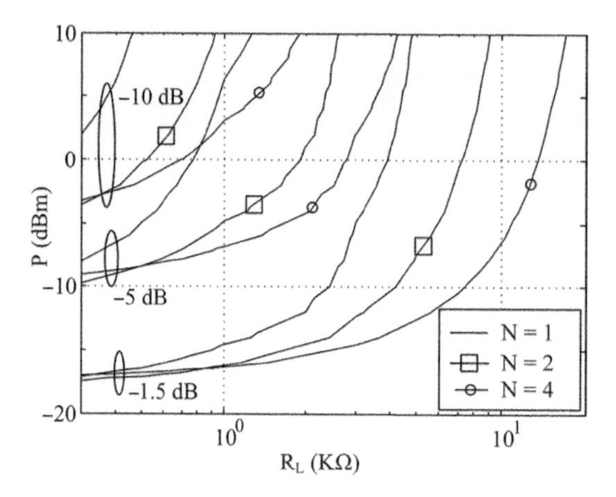

Figure 7.15 Contours of theoretical minimum reflection coefficient magnitude over input power and load resistance for different rectifier circuits [216].

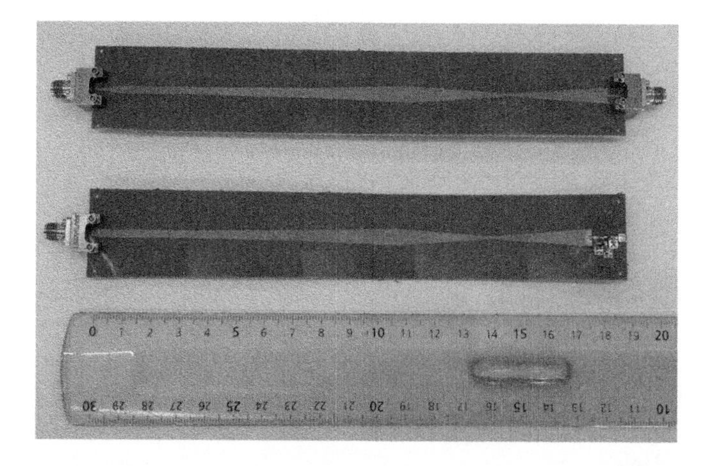

Figure 7.16 Octave band rectifier using a nonuniform transmission line matching network: transmission line prototype (top) and transmission line with rectifier (bottom) [216].

segments of fixed legth and width, hence impedance, and then optimizing the width and the number of sections while placing minimum efficiency goals at different frequency points covering the desired frequency band. The optimization process was done for an input powet of −20 dBm. A series inductor was placed at the input of the rectifier to facilitate the optimization process. The resulting rectifier is shown in Figure 7.16, while its input impedance and RF-dc conversion efficiency are shown in Figures 7.17 and 7.18 respectively.

As expected, the input matching is improved as the input power increases because the rectifier input resistance is reduced to values closer to 50 ohms (Figure 7.18). The input impedance and resulting efficiency are optimized over an octave bandwidth. One disadvantage of a nonuniform transmission line is its large

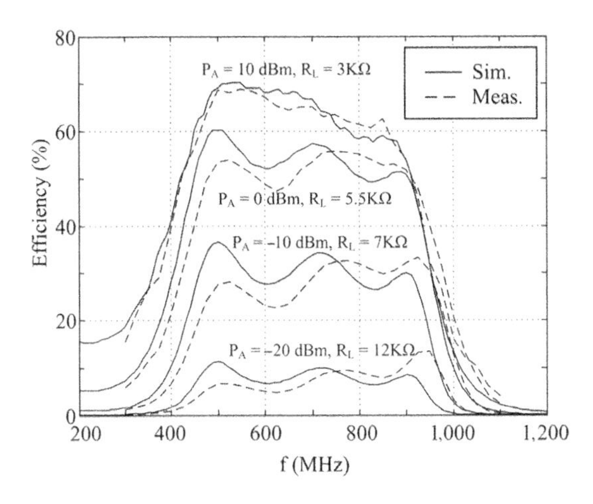

Figure 7.17 Efficiency of ultrawideband rectifier using a nonuniform transmission line matching network [216].

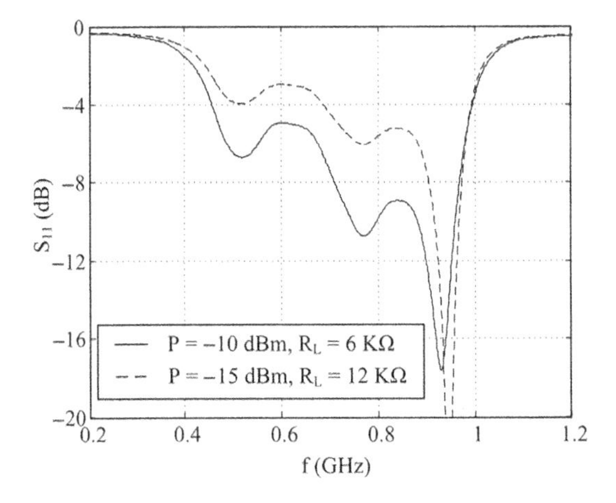

Figure 7.18 Measured input s-parameters of ultrawideband rectifier using a non-uniform transmission line matching network [216].

size; however, meandering can help reduce its overall layout area. Alternatively, one may approximate a nonuniform line using cascaded series inductor L and shunt capacitor C sections [218].

7.7 Load Resistance and Input Power Effects on Rectifier Efficiency

We have seen in Section 7.4 that the rectifier efficiency depends strongly both on the output load resistance and on the input power. However, it is reasonable to consider variable load conditions at rectifier circuits depending on the power requirements of the circuits following the rectifier. Therefore, reduced sensitivity

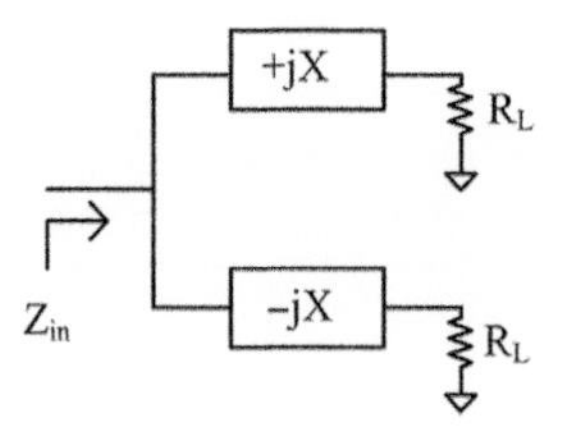

Figure 7.19 Resistance compression network [219].

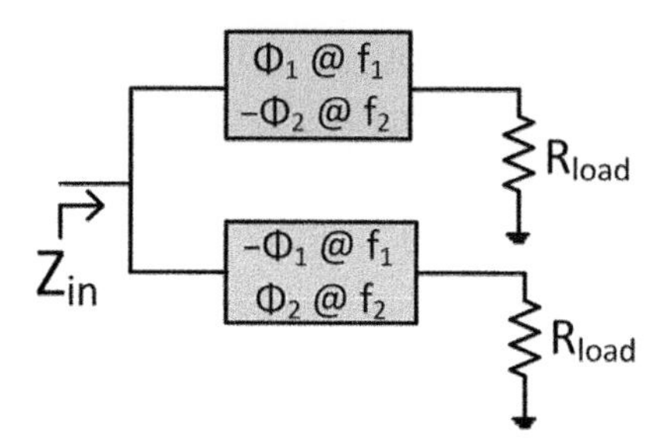

Figure 7.20 Dual-band resistance compression network. ©2014 IEEE. Reprinted with permission from [220]

to load variations is a desirable property for a rectifier circuit. One possible way to achieve this is using a resistance compression circuit [219]. An example of a resistance compression circuit proposed by [219] is shown in Figure 7.19. It comprises of two branches that are connected in parallel and they are each terminated to identical variable resistive loads R. Reactance components with opposite values $\pm X$ are connected in series with the resistive loads. A straightforward analysis shows [219]

$$R_{in} = \frac{X^2}{2R}\left[1 + \left(\frac{R}{X}\right)^2\right]. \tag{7.26}$$

The reactance elements of the branches can be implemented, for example, with one series inductor L at one branch and a series capacitor C at the second branch such that at a desired operating frequency $LC\omega_o^2 = 1$. It is easy to verify that a variation in R is translated to a smaller variation at the input of the resistance compression circuit at the expense of a more complex circuit comprising a larger bill of components and two identical loads. It is possible to cascade more than one such network in order to optimize the resistance compression [219].

It is possible to synthesize multiband resistance compression networks by synthesizing reactive networks with opposite reactance values at the desired frequency points. A dual-band topology is pictured in Figure 7.20.

The implemented resistance compression network and the fabricated prototype are shown in Figures 7.21 and 7.22. The operating bands were 915 MHz and 2.45 GHz. The reactive network comprised two cells each comprising a series LC

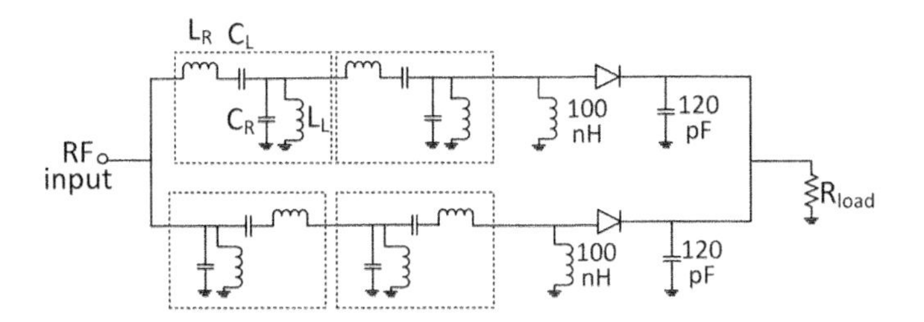

Figure 7.21 Dual-band resistance compression network circuit implementation operating at 915 MHz and 2.45 GHz ($L_R = 8.7$ nH, $L_L = 100$ nH, $C_R = 0.8$ pF, $C_L = 2.7$ pF). ©2014 IEEE. Reprinted with permission from [220]

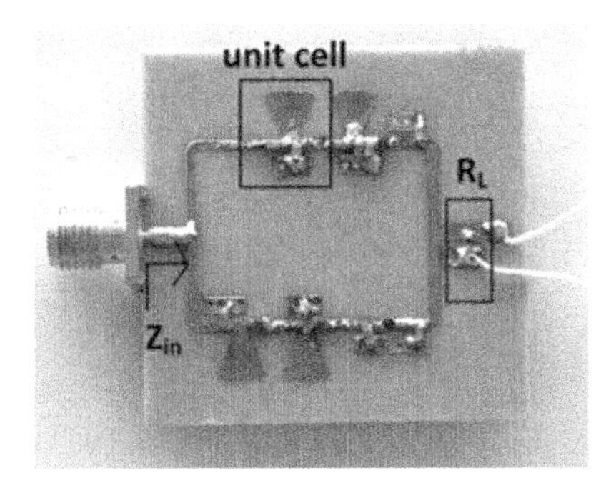

Figure 7.22 Dual-band resistance compression network prototype operating at 915 MHz and 2.45 GHz. ©2014 IEEE. Reprinted with permission from [220]

and a shunt LC circuit providing the desired reactances at the two frequency bands. Two cells were used in order to obtain a better impedance match at the two frequencies. It is interesting to see that one load R_L is used, however, two rectifier circuits are necessary to implement the two branches that are required for the resistance compression network.

Due to the fact that the impedance matching network of the rectifier is designed for a specific load value, load variations result in input impedance variations that lead to impedance mismatch and reflection losses reducing the rectifier efficiency. The dual cell reactive network introduces both the necessary reactance for the resistance compression and transforms the impedance of the rectifier in order to provide impedance matching. The obtained efficiency is compared to a standard dual-band two-diode rectifier without a resistance compression network, shown in Figure 7.23. A plateau in the efficiency is observed at both frequencies for a range of load values demonstrating the effectiveness of the resistance compression.

It turns out that the resistance compression network additionally slightly reduced the sensitivity of the efficiency in variations in the input power levels. The property of the efficiency being insensitive to input power variations is particularly attractive because ambient RF sources have an inherently random

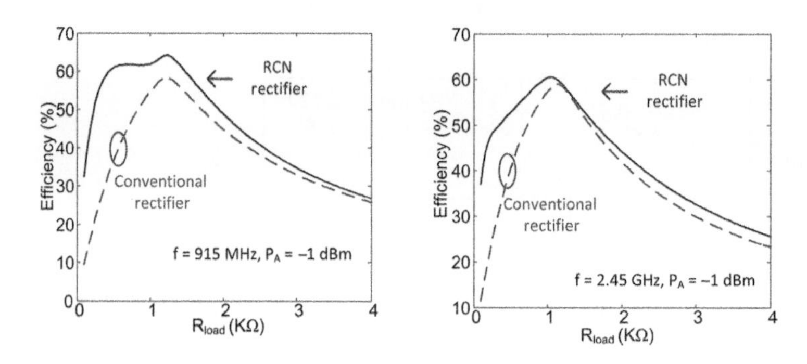

Figure 7.23 Efficiency of a dual-band resistance compression network [220].

and time varying nature due to propagation effects but also modulation present in the RF signals. One way to reduce the sensitivity to input power variation that has been proposed in [221] is to implement a composite rectifier comprising a number of additional diodes and other switching transistor devices that are connected in series with a single shunt diode. The single diode rectifier has a high efficiency at low input power level, but its efficiency drops as the input power increases beyond the point that it drives the diode to its breakdown region. In contrast, a rectifier that contains many diodes in a series has a low efficiency at low input power levels, but it can tolerate a much higher input power before it reaches its breakdown voltage and therefore maintains a high efficiency at higher input power levels than the single diode rectifier. A FET switch is connected in parallel to the additional diodes in such a way that it is at its "On" state, shorting the additional diodes at low input power levels and allowing only the single diode to operate. When the input power increases beyond a critical level, the FET is switched to the "Off" state and the additional diodes are introduced in the rectifier circuit, resulting in a high efficiency for a larger range of input power values.

7.8 Rectification and Angle of Arrival of Incoming Waves

When it comes to harvesting ambient RF energy, the location of the source of the RF energy is typically unknown to the harvester. Consequently, it is necessary to use in the rectenna device an antenna that does not have a directional radiation pattern; rather, an approximately spherical or omnidirectional pattern is desirable. Naturally, in order to maximize the harvested power, it is desirable to form arrays of rectennas, such as [132, 145]. Increasing the number of antennas increases the available effective area for the collection of RF energy, and thus the total RF energy that is being harvested is multiplied by the number of antenna elements. The question arises as to whether it is preferable to sum the harvested energy at the RF level or at the dc level. RF power combining results in forming a directional radiation pattern, therefore sacrificing the omnidirectional characteristic of the harvester. dc power combining maintains the omnidirectional characteristic of the antennas as each antenna is considered individually (with some loading effect from the unavoidable mutual coupling with its neighboring elements). The dc output of each rectenna element can be combined in series or parallel, resulting in a desired output voltage and efficiency [145]. The overall efficiency depends, however, on the loading that each rectenna presents to the rest, and it is affected by variations in dc output to the individual elements due to tolerances or, more importantly, due to variations in the RF power illuminating each array element [222, 223].

As we have seen, however, the RF-dc conversion efficiency of the rectifier depends on the input power, and specifically at low input power levels it increases with the input powet. Therefore, one is tempted to perform some RF combining before converting to dc. As a result, one might consider rectenna systems where one rectifier is connected to an antenna subarray and have each subarray radiation pattern point to a different direction, thereby achieving the superposition of RF energy before the RF-to-dc conversion and covering a wide angular area [224]. Another way around the angular coverage problem could be to employ beam-steering architectures, but such systems consume energy in steering the antenna beam, and therefore, one needs to take into account the amount of energy spent in the beam-steering process in order to compute the overall RF-to-dc conversion efficiency of the system.

An alternative rectifier topology based on a Wilkinson power combiner was proposed in [225]. It comprises a two-branch rectifier, which is capable of maintaining an approximately constant RF-to-dc conversion efficiency over any arbitrary phase shift between the RF signals present at its input terminals and, therefore, over any angular direction. Furthermore, it is scalable, employing combiner modules connected to a large number of antenna elements or subarrays and subsequently combining the output in series or parallel configuration. The circuit performs both RF and dc power combining and is shown in Figure 7.24. It comprises two inputs that would be connected to the antenna elements. A Wilkinson combiner combines the two inputs, and the combined signal is rectified by a first series diode rectifier circuit. The Wilkinson combiner circuit includes a resistor connected across its two inputs, which helps isolate the two inputs. When the two inputs are in phase, there is no current flowing through the resistor, and the two inputs are combined at the output port. When, however, there is a phase difference at its inputs, there is some current flowing in the resistor that helps maintain input matching and isolation at the expense of the power dissipated in the resistor. In the topology of Figure 7.24, a second rectifier is used in place of the resistor, which maintains the input matching of the ports but additionally converts (a fraction) of the RF power flowing into dc power and minimizes the dissipated RF power. The two rectifier outputs are summed in parallel in order to obtain a single dc output.

The input matching networks of the rectifiers are optimized in such a way that the efficiency of the rectifier remains relatively constant independently

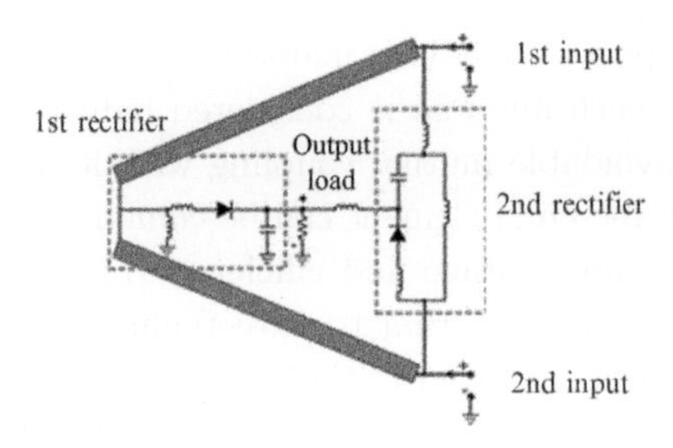

Figure 7.24 Circuit schematic of RF and dc combining rectifier based on a Wilkinson power combiner that is insensitive to phase differences at its inputs. ©2019 IEEE. Reprinted with permission from [225]

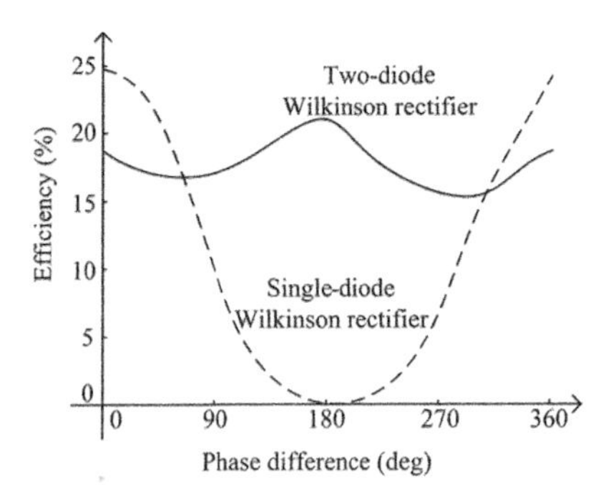

Figure 7.25 RF and dc combining rectifier efficiency versus the phase difference at its input terminals for an input power of −20 dBm at each RF input port [225].

of the phase difference between the two RF input signals. As a result, the circuit achieves both a higher antenna gain due to the RF power combining and maintains a high dc combining efficiency. Because of the fact that the rectifier is a nonlinear circuit, its efficiency increases nonlinearly with the input power. Consequently, because the RF combining results in a higher RF input power to the rectifiers, it is possible to operate at a higher RF-dc conversion efficiency than a topology where each antenna is individually connected to a rectifier without performing any RF power combining. The rectifier efficiency is plotted in Figure 7.25 versus the phase difference, compared to a Wilkinson-based rectifier with a single two-diode rectifier at the power-combining output of the Wilkinson power divider, where the shunt resistor is maintained. The prototypes of the two circuits that were used to obtain the measured results of Figure 7.25 are shown in Figure 7.26. Using a rectifier in place of the shunt resistor allows one to maintain good efficiency for any phase difference between the input RF signals.

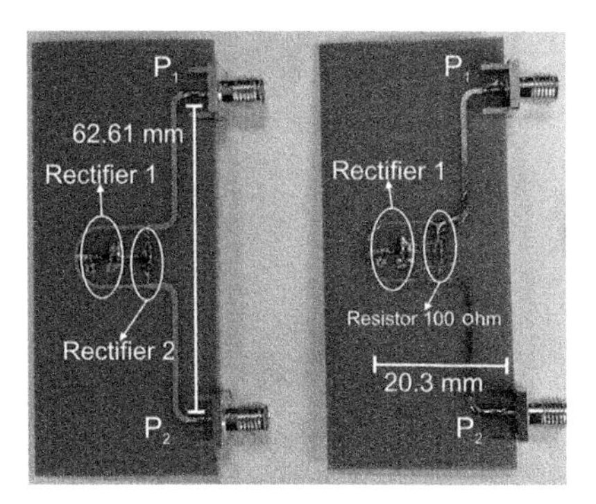

Figure 7.26 RF and dc combining rectifier prototypes. ©2019 IEEE. Reprinted with permission from [225]

7.9 Signal Optimization for RF Energy Harvesting

Wireless power transmission is associated with transferring microwave power using a continuous wave (CW) carrier signal. Ambient RF signals, however, are not necessarily continuous wave signals. In fact, wireless communication signals are modulated signals that typically have a finite frequency bandwidth and a time varying envelope. It is therefore natural to ask the question of whether the properties of modulated signals affect the power transfer efficiency or, alternatively, what are the characteristics of an optimum signal for wireless power transmission.

Measurements of RF signals with a time-varying envelope have appeared in the literature to the best of our knowledge as early as 2005 [132], where measurements with a two-tone composite RF signal were performed, while later in 2010 measurements of RF-dc power conversion using a quadrature phase shift keying (QPSK) signal showed a small efficiency improvement relative to a CW tone signal [226].

We can describe such signals with a time-varying envelope using a complementary cumulative distribution function (CCDF) plot. In the case of randomly varying signals such as digitally modulated signals, the CCDF curve $\overline{F}_p(p)$ represents the probability Pr that a random variable, in this case the instantaneous power P of a signal, has a certain value that is larger than a value p [227],

$$\bar{F}_p(p) = \Pr\{P > p\}. \tag{7.27}$$

The CCDF is related to the cumulative distribution function (CDF), $F_p(p) = \Pr\{P < p\}$ as $\bar{F}_p(p) = 1 - F_p(p)$. The ratio of the instantaneous power over the average power in dB is the abscissa of a CCDF plot [228]. In the case of deterministic periodic signals such as multitone signals (also called multisine signals), the CCDF curve represents the fraction of the envelope period that the instantaneous power takes a desired value that is larger than the average. As an example, CCDF curves of multitone signals with two, three, four, and eight

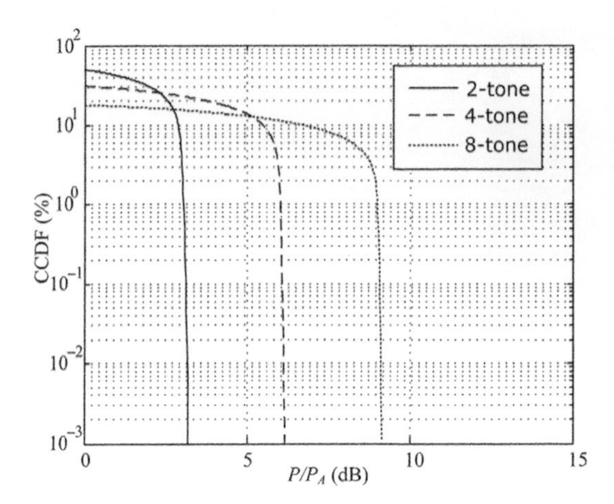

Figure 7.27 Measured CCDF curves of multitone signals with two, four, and eight tone carriers [209].

carriers is shown in Figure 7.27 [209]. A carrier spacing of 0.5 MHz was used around an RF carrier value of 915 MHz. The measurements were performed using a Keysight ESG 4438C Digital vector signal generator with multitone signal generation capability and a Vector Signal Analyzer (VSA) running on a PSA E4448A Spectrum Analyzer as a receiver. The CCDF curves have a shape resembling a waterfall curve. The rightmost point of the curve represents the peak-to-average power ratio (PAPR) value of the signal. Usually, for practical reasons, a specific value of CCDF probability (or time fraction) such as 0.1% is used to determine the PAPR value. One can verify that as the number of tones increases, the PAPR value also increases. It should be emphasized, however, that the PAPR and CCDF curves in general depend on the relative phase value between the carriers. In this case, all tones were synchronized in phase with each other.

Selected measured CCDF curves of digitally modulated signals are shown in Figure 7.28 [209]. In digitally modulated signals, a pulse-shaping filter is used in order to limit the bandwidth of the transmitted waveforms [229]. A commonly used filter is a square root cosine filter, which is characterized by the roll-off factor $0 \leq \beta \leq 1$. The CCDF curves were created using a roll-off factor $\beta = 0$, which limits the bandwidth of the transmitted signals to f_s, where f_s is the symbol rate. The roll-off factor affects the PAPR, as shown in Figure 7.29.

Until this section, we have seen that both the average input power and the output load affect the RF-dc conversion efficiency of a rectifier. In Figure 7.30, we created contours corresponding to an RF-dc conversion efficiency of 20% versus the available average input power P_A and the output load R_L of a series rectifier [209]. The contours appear to be suddenly interrupted and overlap each other at high load and high power values due to the fact that the diode breakdown voltage is reached and efficiency suddenly drops. One important conclusion that can be drawn from these contours is that as the PAPR of the signal increases the efficiency contours are shifted toward higher load values. This is an important

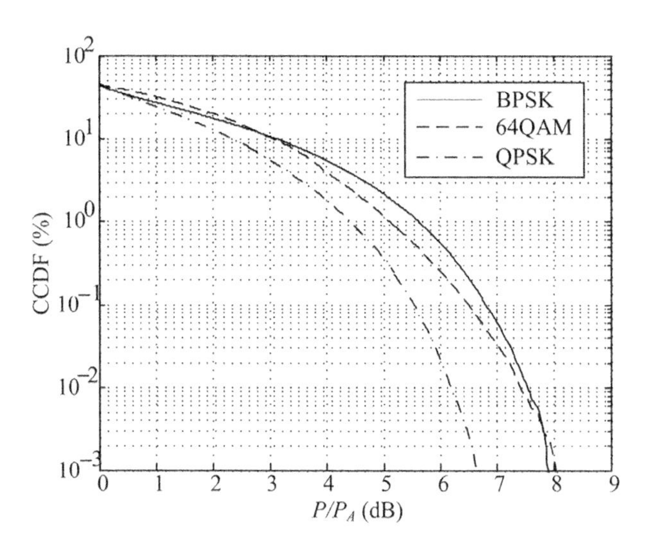

Figure 7.28 Measured CCDF curves of digitally modulated signals with pulse-shaping roll-off factor $\beta = 0$ [209].

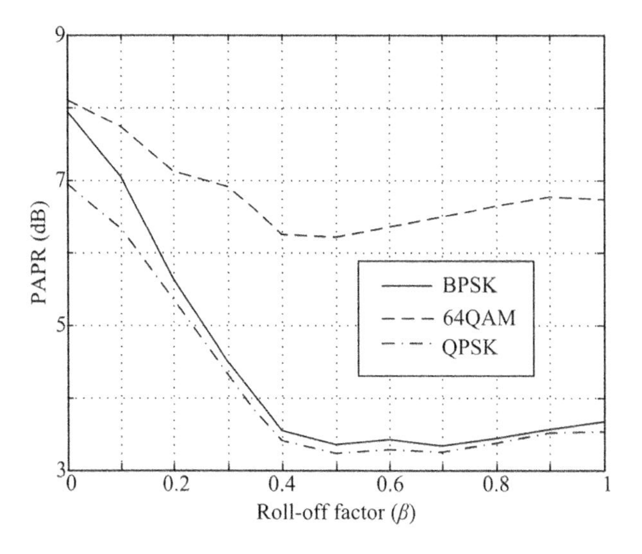

Figure 7.29 Measured PAPR of digitally modulated signals versus the pulse-shaping roll-off factor β [209].

result because it shows that one can control the efficiency by selecting a signal with a certain PAPR. Results demonstrating that a higher RF-dc conversion efficiency can be obtained using multitone signals have been presented in [147] and later in [202]. The multitone signals were described as power-optimized waveforms and a theoretical analysis of the RF-dc conversion efficiency was presented in [230].

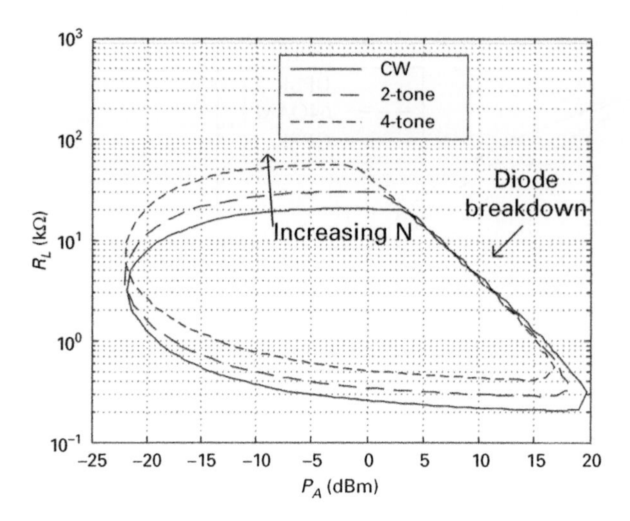

Figure 7.30 Contours of 20% RF-dc conversion effficiency of different multitone signals applied to a series diode rectifier [209].

A similar behavior occurs for randomly modulated signals with a time-varying envelope. Measurements comparing the RF-dc conversion efficiency of a rectifier to which we applied different modulated waveforms of the same bandwidth but with different PAPR were first presented in [149]. The different waveforms where an orthogonal frequency-division multiplexing (OFDM) signal, white noise and a chaotic signal generated from a properly biased Colpitts oscillator [149]. Each of these signals has a different, increasing PAPR value that was determined experimentally, as shown in Figure 7.31. The RF-dc conversion efficiency measured over a 5.6 KΩ load is plot in Figure 7.32. Similarly with the multitone signals, we observed that a higher conversion efficiency is obtained at low input power levels for signals with a higher PAPR. Once the power increases, the signals with a high PAPR first result in the rectifier reaching the breakdown voltage of the diode and their efficiency begins to drop. It should be emphasized, however, that, as Figure 7.30 showed, the efficiency contours shift toward higher load resistance values with increasing PAPR, and therefore the observed efficiency improvement is contingent on the load resistance that has been used. However, high load resistance values are typically associated with low power sensors because of the inherent low power dissipation requirement, and therefore having a high efficiency at high load values appears to be an attractive property.

The PAPR value, although important, does not fully describe the rich nature of the signal envelope variations. For example, we experimentally synthesized two signals with the same PAPR but otherwise having a different CCDF curve behavior. These two signals were a four-tone signal with in-phase tones and a 64QAM signal with a roll-off $\beta = 0.5$. The measured CCDF of the signals is shown in Figure 7.33. Although the two signals have the same PAPR, the four-tone signal has a wider CCDF curve, implying a stronger instantaneous power

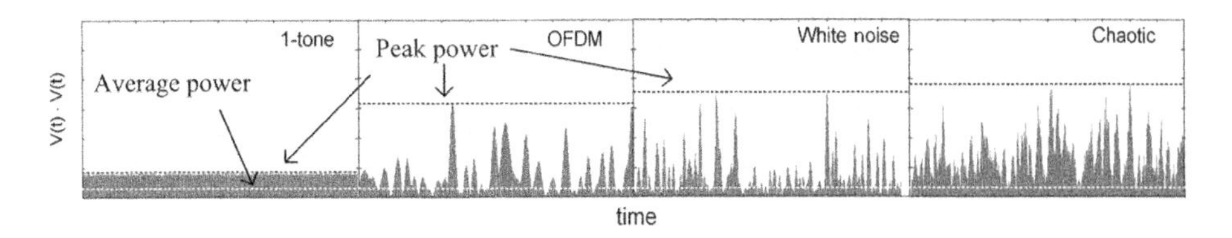

Figure 7.31 Instantaneous power of different randomly modulated signals with a time-varying envelope and different PAPR values. ©2014 IEEE. Reprinted with permission from [149]

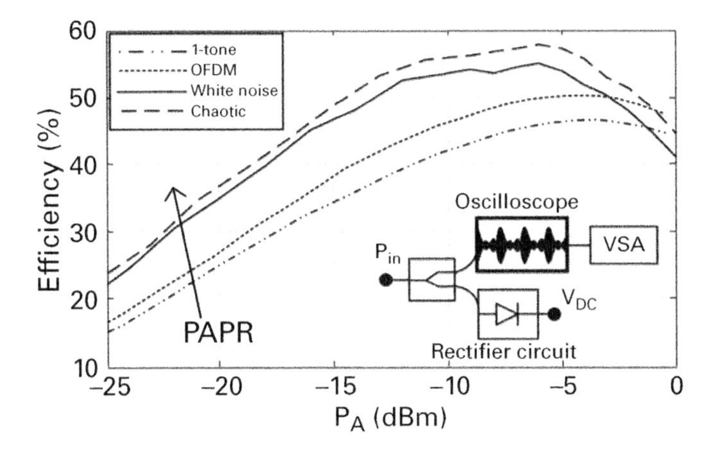

Figure 7.32 RF-dc conversion effficiency of different randomly modulated signals with a time-varying envelope and different PAPR values. ©2014 IEEE. Reprinted with permission from [149]

variation where the instantaneous power of the four-tone signal takes values closer to its peak value more times than the 64 QAM signal. We then applied the signals to a rectifier and measured the RF-dc rectifier efficiency, shown in Figure 7.34.

One can see that even though the two signals have the same PAPR, the obtained efficiency curves are very different. Specifically, the efficiency of the four-tone signal is shifted toward higher load values. It is possible to distinguish between these two signals by using a different parameter, the instantaneous power variance [231]. The instantaneous power variance is defined as

$$\sigma_P^2 = \mathrm{E}\left[(P - P_A)^2\right] = \int_0^{+\infty} (P - P_A)^2 dF_p = -\int_0^{+\infty} (P - P_A)^2 d\bar{F}_p, \quad (7.28)$$

where E[] denotes expectation and P_A is the average power value,

$$P_A = \int_0^{+\infty} P dF_p = -\int_0^{+\infty} P d\bar{F}_p. \quad (7.29)$$

The instantaneous power variance measures the "width" of the CCDF curve and describes in a sense more accurately than the PAPR the envelope of the signals.

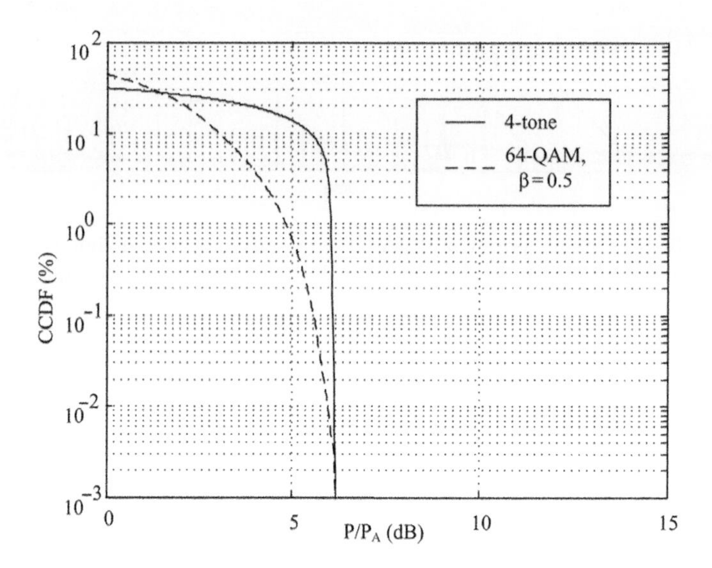

Figure 7.33 CCDF comparison of a four-tone signal and a 64QAM signal with $\beta = 0.5$ with the same PAPR $\approx$ 7dB. ©2016 IEEE. Reprinted with permission from [209]

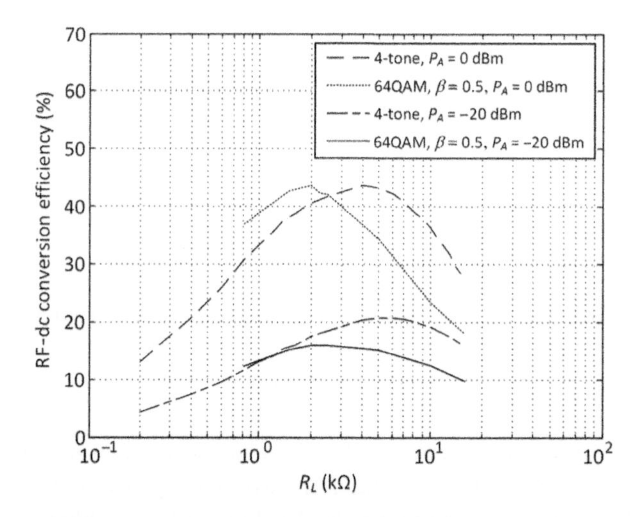

Figure 7.34 RF-dc efficiency comparison obtained using the signals of Figure 7.33. ©2016 IEEE. Reprinted with permission from [209]

It is possible to compute the instantaneous power variance using the CCDF values; however, one should be careful that measurement equipment typically plots the CCDF curve only for values that correspond to power levels that are higher than the average power, and therefore access to the rest of the CCDF values corresponding to power levels below the average power is required. The four-tone signal of Figure 7.33 has a larger instantaneous power variance than the 64 QAM signal and in accordance with the previous measurements it results in the RF-dc conversion efficiency shifting toward higher load values.

7.10 Problems and Questions

1. What is the power density of an electric field with strength $1V/m$?
2. What is the current responsivity of a detector circuit?
3. What is approximately the state-of-the-art diode rectifier efficiency (based on the available input power) in the low GHz frequency range and for low available input power levels around -20 dBm?
4. Starting from (7.22), compute an approximate expression of the input RF resistance of a series diode rectifier to the limit of small input voltage V_{11}.
5. What is (are) the reason(s) that the efficiency of a diode rectifier is reduced at high input power levels?
6. Consider a charge pump rectifier with $N = 4$ diodes according to Figure 7.14. What is the theoretical maximum bandwidth that can be achieved with a maximum input reflection coefficient magnitude of 0.1 for an available input power of -20 dBm and -10 dBm?
7. What is a resistance compression network? Compute the required inductance L and capacitance C of a resistance compression network operating at 0.9 GHz, where the load resistance varies from 25 Ω to 100 Ω and the input resistance only varies between a value R_{in} and $2R_{in}$.
8. What is the PAPR ratio and the instantaneous power variance of a signal? What can they tell us in terms of the performance of a signal in RF energy harvesting?
9. Based on Figure 7.29, how do BPSK, 8PSK, and QPSK with a roll-off factor $\beta > 0.4$ compare in terms of their performance as RF energy harvesters? Similarly, based on Figure 7.33 how do 64QAM and a four-tone signal compare in terms of PAPR and instantaneous power variance and in terms of their performance as RF energy harvesters?

8 Power Supplies and Storage

8.1 Introduction

Electronic circuits are designed to operate with certain voltage supply levels, ranging typically from 0.9 V to 5 V for low-power circuits related to wireless communication and sensing applications. Therefore, the output of an energy harvesting device must be converted to a suitable supply voltage in order to power the electronics connected to it. Power converter circuits are necessary both in order to convert the energy harvester output to a desired useful voltage value but also to regulate the output voltage of the energy harvester to a constant value insensitive to variations. Power converter circuits are classified into regulated or unregulated depending on whether they have functionality to maintain a constant output voltage and attenuate any ac ripple or variation at their input. One characteristic measure of a regulator is the line regulation (LNR) expressed as the ratio of the output voltage variation ΔV_L to a corresponding change in the input voltage ΔV_i at a constant output current I_L and temperature T_A,

$$\mathrm{LNR} = \left.\frac{\Delta V_L}{\Delta V_i}\right|_{I_L, T_A = \mathrm{const.}} \tag{8.1}$$

The line regulation is typically measured in mV/V.

When the input power to the harvester device is very small, it is often the case that the output voltage of the energy harvester is less than the desired operating voltage. This is the case, for example, of an RF energy harvester, a rectenna comprising a single-diode rectifier when the input power to the rectifier is in the order of −20 dBm or less (see Section 7.4). Alternatively, certain types of energy harvesters such as electrostatic or piezoelectric energy harvesters (see Section 4.6) provide very high voltage values at their output that need to be converted to a lower value suitable for the subsequent electronic circuits. Power converter circuits are also classified as step-down or step-up, depending on whether they provide an output voltage that has a value that is less or higher than the voltage applied at its input respectively. Power converter circuits may have both a step-down and step-up functionality, as we will see in the next sections.

Finally, power converter circuits are classified into linear or switched mode depending on their operating principle. There are other ways to classify power converters, and furthermore, both linear and switching power converters can be

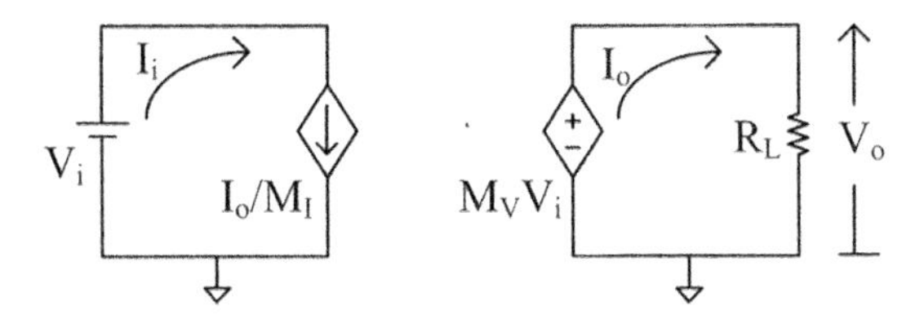

Figure 8.1 Simple circuit model of a dc–dc power converter.

classified into subcategories based on their topology and functionality [232]. One of the most important characteristics of the dc–dc power converter circuits is the efficiency η defined as

$$\eta = \frac{P_L}{P_i} = \frac{P_L}{P_L + P_d}, \tag{8.2}$$

where P_L is the dc power delivered to a load R_L connected at its output and P_i is the input power to the power converter circuits, which, in contrast to the energy harvester circuits of the previous chapters, is also a dc electrical power same as the output power P_L. P_d is the dissipated power of the dc–dc converter circuit given by

$$P_d = \left(\frac{1}{\eta} - 1\right) P_L. \tag{8.3}$$

The dc voltage and current gain of a dc–dc converter are defined as

$$M_V = \frac{V_L}{V_i} \tag{8.4}$$

$$M_I = \frac{I_L}{I_i}. \tag{8.5}$$

Using the preceding gain expressions, the efficiency can be expressed as

$$\eta = M_V M_I. \tag{8.6}$$

This way, one can define a dc circuit model of a dc–dc power converter as shown in Figure 8.1 [232].

The input dc resistance of the switched mode converter is defined as

$$R_i = \frac{V_i}{I_i}. \tag{8.7}$$

Using (8.4) through (8.6), one can write

$$R_i = R_L \frac{\eta}{M_V^2} \approx \frac{R_L}{M_V^2}. \tag{8.8}$$

An ideal switched mode converter has an efficiency of 100%, and therefore, using (8.8), we can find the input resistance of a switched mode converter based on the output load and the voltage gain.

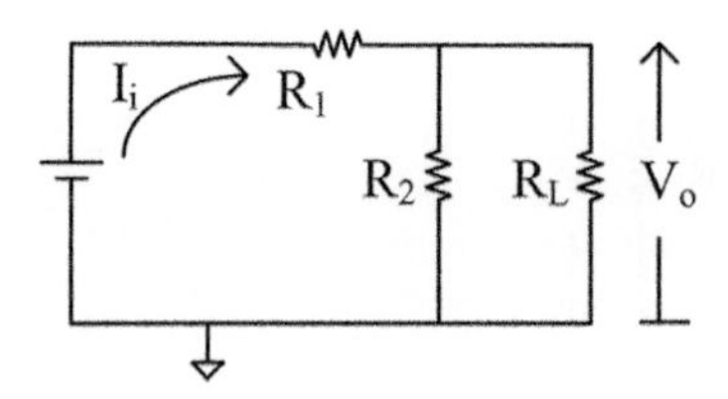

Figure 8.2 Step-down converter principle using a resistive voltage divider circuit.

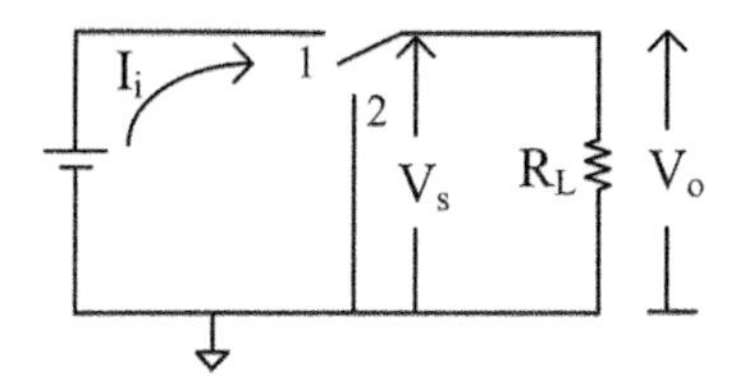

Figure 8.3 Step-down converter principle using an SPDT switch [233].

The simplest type of a step-down dc–dc power converter circuit is a resistive voltage divider circuit, shown in Figure 8.2. It is well known that the load voltage V_L is related to the input voltage by the simple formula

$$V_L = \frac{R_L /\!/ R_2}{R1 + R_L /\!/ R_2} V_i. \tag{8.9}$$

The circuit of Figure 8.2, albeit simple, typically results in low efficiency due to the power dissipated in resistors R_1 and R_2. Alternatively, one can produce a lower dc value using the circuit topology of circuit (8.3) [233]. A single-pole double-throw (SPDT) switch is used to connect the output load R_L for a fraction D of a period T_s of a control voltage v_s to the input voltage V_i, while it shorts the load to the ground for the remaining fraction $1 - D$ of the period of v_s. It is easy to verify that the average (dc) output voltage is equal to

$$V_L = \frac{1}{T_s} \int_0^{T_s} v_L(t)dt = DV_i. \tag{8.10}$$

A switch device connects, in a closed position, or disconnects, in an open position, a galvanic electrical path in a circuit. In its simple form, it has two contacts or terminals. Fundamental electrical switch components are a diode and a transistor device. A control signal is necessary to set the switch to the closed or open state. In the case of the diode, the control signal is the voltage, which is applied to its terminals, whereas in a transistor device it can be a voltage or a current applied to its third terminal, depending on the transistor technology. The number of poles represents how many switch circuits are controlled simultaneously by the control signal. The number of throws defines how many different closed paths are adopted by the switch. In the case of Figure 8.3, there is one circuit, hence a single pole switch is used, and furthermore there are two closed paths involved, hence the switch is a double throw switch. The elementary diode

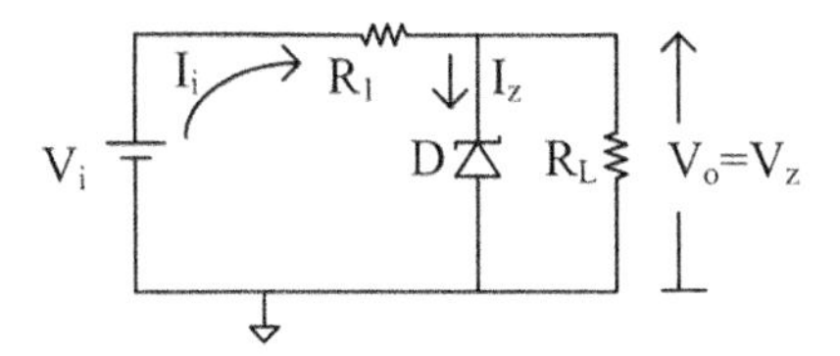

Figure 8.4 Step-down converter using a Zener diode.

and transistor components define single-pole single-throw (SPST) switches. An SPDT switch can be constructed from two elementary SPST switches that are controlled by a single control signal. As we will see in Section 8.3, the SPDT switches of switched mode dc–dc power converters are commonly implemented using a combination of a transistor and a diode device.

If there is no power dissipated in the switch, one can achieve a very high efficiency switched mode dc–dc power converter. This is done at the expense of an increased circuit complexity, in terms of the control signal circuitry and filtering circuitry in order to attenuate signal components created by the switching process at the switching frequency harmonics. One however must also consider the power lost in implementing the switching signal, which lowers the efficiency and could be important in implementing energy harvesting circuits.

8.2 Linear Power Converters

Linear power converters are step-down power converters. The simplest form of a regulated step-down dc–dc power converter is one using a Zener diode in place of the resistor R_2 of Figure 8.2, as shown in Figure 8.4. When the input voltage is larger than the Zener breakdown voltage V_Z, the output voltage V_L remains constant at V_Z. It therefore forms a regulated step-down supply. Naturally, all the power that is dissipated in the Zener diode $P_d = V_Z I_Z$ is lost. The efficiency of the supply is reduced as the difference between the input and output voltage is increased.

The Zener diode represents a variable resistor connected in parallel to the output load. The variation of the resistance of the Zener diode results in the regulation of the output load voltage. Alternatively, one can implement a variable resistor using a transistor device that is biased in its active region [232, 233]. The resulting circuit is a linear regulator circuit. A bipolar transistor biased in its active linear region can be used in place of either the resistor R_1 or R_2, resulting in a series or shunt linear regulator topology as shown in Figure 8.5 [232].

In Figure 8.5, a feedback loop is created using a voltage divider comprising resistors R_1 and R_2 which sense the output voltage and compare it to a reference value V_{ref} with the help of an operational amplifier. The output of the amplifier drives the base of the bipolar transistor, thus controlling its collector current. This way, the collector to emitter resistance of the transistor changes according

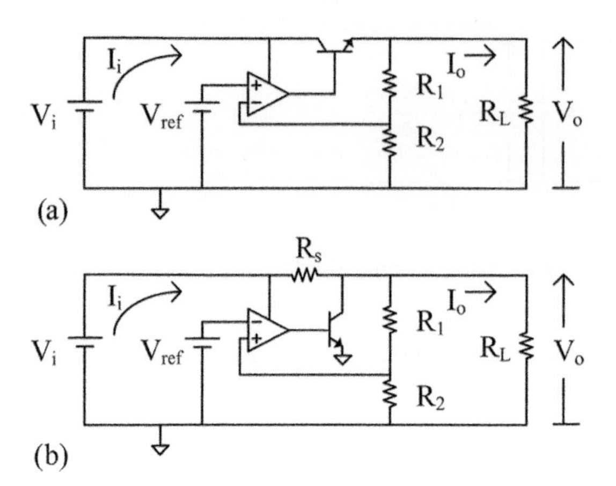

Figure 8.5 Linear voltage regulators: (a) series and (b) shunt topologies.

to the output voltage, thus maintaining a regulated output voltage V_L. In the case of the series linear regulator, the output voltage V_L is calculated as follows:

$$V_{R_2} = \frac{R_2}{R_1 + R_2} \approx V_{ref} \tag{8.11}$$

$$V_L = \left(\frac{R_1}{R_2} + 1\right) V_{ref}. \tag{8.12}$$

The output voltage is determined by the feedback resistive divider and the reference voltage. Furthermore, the efficiency of the linear regulator is easily calculated using the fact that $I_I \approx I_L$ as

$$\eta = \frac{P_L}{P_I} = \frac{V_L I_L}{V_i I_i} \approx M_V. \tag{8.13}$$

The closer the output voltage V_L is to the input voltage V_i, the higher the efficiency is. However, the efficiency can never become 100% due to the fact that there is always a small voltage drop V_{CE} associated with the transistor terminals. The minimum voltage difference between the input and the output voltage of the regulator is called the dropout voltage. The dropout voltage typically takes a value of 2V; however, there also exist low dropout (LDO) linear regulators with a dropout voltage as low as 0.1 V [232].

It is straightforward to compute the output voltage and the efficiency of the shunt linear regulator with similar circuit equations. The efficiency of the shunt linear regulator is less than the efficiency of the series linear regulator due to the fact that in addition to the transistor there is also power dissipated in resistor R_1 [232].

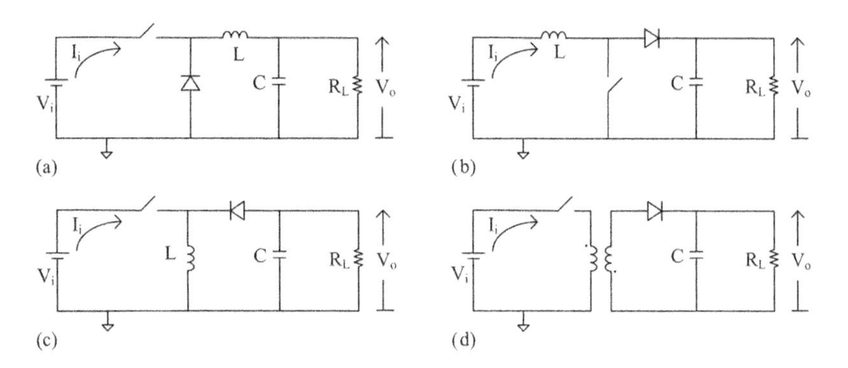

Figure 8.6 Switched mode power converter topologies: (a) buck, (b) boost, (c) buck-boost, and (d) flyback.

8.3 Switched Mode Power Converters

There exist a number of different switched mode power converter topologies. Some of the most common ones are the buck, boost, buck-boost, and flyback converter, all shown in Figure 8.6. The buck converter is a step-down converter, and the boost converter is a step-up converter, whereas the buck-boost and the flyback converters are step-up/down converters. The switch element in Figure 8.6 is an SPST switch comprising a transistor controlled by a periodic switching waveform. Combined with the second SPST switch comprising the diode, they implement the SPDT switch required for the converter circuit. The flyback converter is a buck-boost converter where the inductor has been substituted by a transformer. One should further note that the output voltage of the buck-boost converter has opposite polarity to the input voltage. There exist many more switched mode converter topologies [232, 233].

8.3.1 Steady-State Analysis

The steady state of the switched mode power converter circuits can be approximately analyzed employing three principles [233]. The first principle is the small ripple approximation, where it is assumed that under steady-state conditions, the ripple of the current and voltage at the circuit output node are small compared to the dc current or voltage and are ignored in the analysis. In other words,

$$v_L(t) = V_L + u_l(t) \approx V_L, \tag{8.14}$$

where V_L represents the average dc component and $v_l(t)$ the ac ripple component of the total output voltage $v_L(t)$. The second principle is that of inductor flux linkage balance or volt-second balance. This principle states that the average voltage through the inductor is equal to zero. This assumption is correct assuming an ideal inductor, where there are no thermal dissipation losses:

$$< v_L(t) >= \frac{1}{T_s}\int_0^{T_s} v_L(t)dt = 0. \tag{8.15}$$

The third assumption is the capacitor charge balance or ampere second balance, which defines that the average current through the capacitor is equal to zero:

$$< i_C(t) >= \frac{1}{T_s}\int_0^{T_s} i_C(t)dt = 0. \tag{8.16}$$

Applying these three principles, one can compute an estimate of important circuit parameters of the various switched mode converters under steady-state conditions, such as the voltage gain and efficiency. Furthermore, one can estimate the dc input resistance of the converter, which is essential when interfacing an energy harvester circuit such as a rectenna with a switched mode dc–dc converter in order for it to present an optimum load to the harvester circuit ensuring maximum power transfer. The steady state of switched mode power converter circuits is classified into two modes: the continuous conduction mode (CCM) and the discontinuous conduction mode (DCM) . In the former, the current in the magnetic energy storage component (inductor or transformer) is always nonzero, whereas in the latter it goes to zero for a fraction of the switching period T_s. The two modes result in substantial differences in the performance of the converter circuits, and they can be analyzed using the principles defined in this section. In the next sections, we will study the boost converter in detail and present a summary of the voltage gain and input resistance of the different converter topologies.

8.3.2 The Boost Converter

Let us analyze the steady state of the boost converter, shown again for completeness in Figure 8.7 [232, 233]. In the continuous conduction mode, the circuit takes one of two possible configurations based on the position of the switch, shown in Figure 8.8

Let us further assume that the switch is in the first position for a fraction D of the switching period T_s while it remains in the second position for a fraction $D_1 = 1 - D$. In the first position, the inductor voltage and the capacitor current are

$$\begin{aligned} v_L &= V_i \\ i_C &= -\frac{V_L}{R_L}, \end{aligned} \tag{8.17}$$

where the small ripple approximation has been invoked for the output voltage V_L according to (8.14). Similarly, during the second position of the switch, the inductor voltage and the capacitor current become

$$\begin{aligned} v_L &= V_i - V_L \\ i_C &= I_L - \frac{V_L}{R_L}. \end{aligned} \tag{8.18}$$

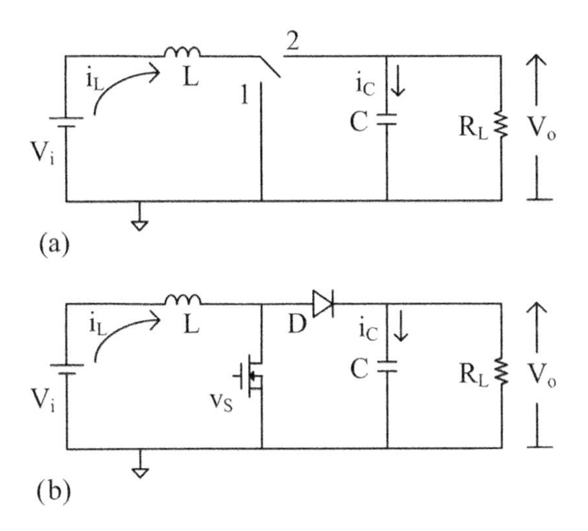

Figure 8.7 Boost converter circuit using (a) an ideal switch and (b) a switch implemented by a power transistor and a diode.

If we invoke the volt-second balance for the inductor L, we have

$$\frac{1}{T_s} = \int_0^{T_s} v_L(t)dt = V_i D + (V_i - V_L)D_1 = 0 \Rightarrow \frac{V_L}{V_i} = M(D) = \frac{1}{1-D}. \quad (8.19)$$

We can then solve for the voltage gain of the boost converter at CCM,

$$M(D) = \frac{V_L}{V_i} = \frac{1}{1-D}, \quad (8.20)$$

which is plotted in Figure 8.9.

Application of the capacitor charge balance gives an expression for the average inductor or output current

$$\frac{1}{T_s} = \int_0^{T_s} i_C(t)dt = -\frac{V_L}{R_L}D + \left(I_L - \frac{V_L}{R_L}\right) D_1 = 0 \Rightarrow I_L = \frac{V_L}{(1-D)R_L}. \quad (8.21)$$

The inductor current is the input current to the converter and therefore we can use it to calculate the input resistance of the boost converter as

$$R_i = \frac{V_i}{I_L} = \frac{(1-D)R_L}{M_V} = (1-D)^2 R_L. \quad (8.22)$$

The inductor L and capacitor C values are selected based on the maximum desirable or allowable current $2\Delta i_L$ (Figure 8.10) or voltage $2\Delta v_C$ (Figure 8.11) variation respectively. In each state, the inductor current has a linear slope that is determined by the equation for the current through the inductor. For example, the current slope in the first state is given by

$$v_L \approx V_i = L\frac{di_L}{dt} = L\frac{2\Delta i_L}{DT_s} \Rightarrow \frac{di_L}{dt} \approx \frac{2\Delta i_L}{DT_s} = \frac{V_i}{L}. \quad (8.23)$$

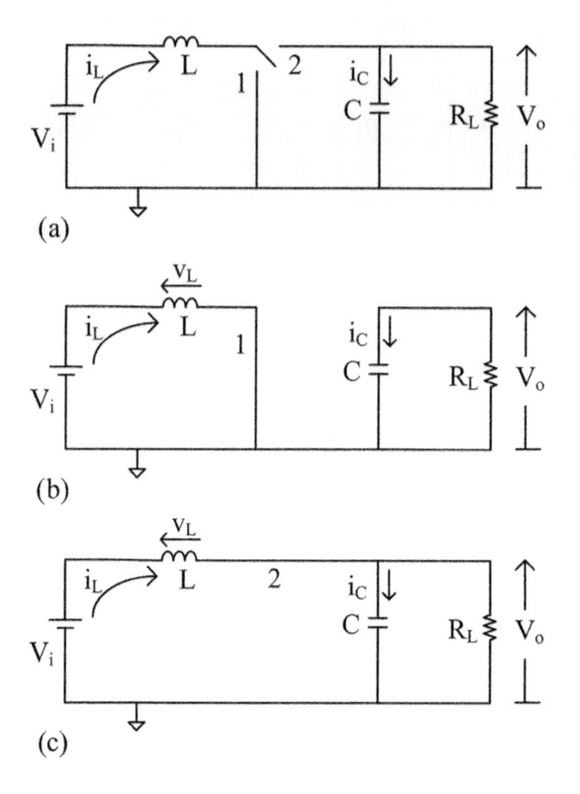

Figure 8.8 Boost converter circuit under CCM operation: (a) converter using an ideal switch, (b) switch in position 1, and (c) switch in position 2.

In the second state, the current slope is

$$\frac{di_L}{dt} \approx \frac{2\Delta i_L}{D_1 T_s} = \frac{V_i - V_L}{L}. \tag{8.24}$$

We can solve (8.23) for the inductance

$$L = \frac{V_i D T_s}{2\Delta i_L}. \tag{8.25}$$

Similarly, the capacitor voltage is linearly increasing and decreasing during the first and second states respectively, and the slope of the voltage curve can be determined from the capacitor current equation (Figure 8.11). The desired capacitor voltage variation Δv_C can be used to determine the capacitance value C. Specifically, during the first position the capacitance current equation takes the form

$$i_C \approx -\frac{V_L}{R_L} = C\frac{dv_C}{dt} = C\frac{-2\Delta v_C}{DT_s} \Rightarrow C = \frac{V_L D T_s}{2R\Delta v_C}, \tag{8.26}$$

which is solved for the capacitance value C.

In the DCM mode, the steady state comprises an additional interval D_2, where the current in the inductor is equal to zero. In this case, the three intervals sum to

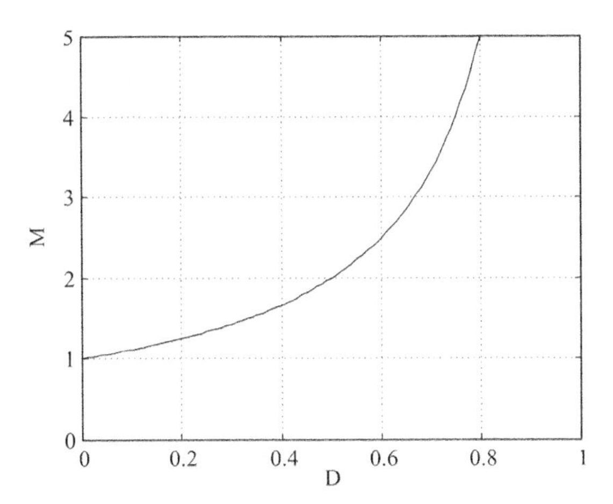

Figure 8.9 Boost converter voltage gain under CCM operation.

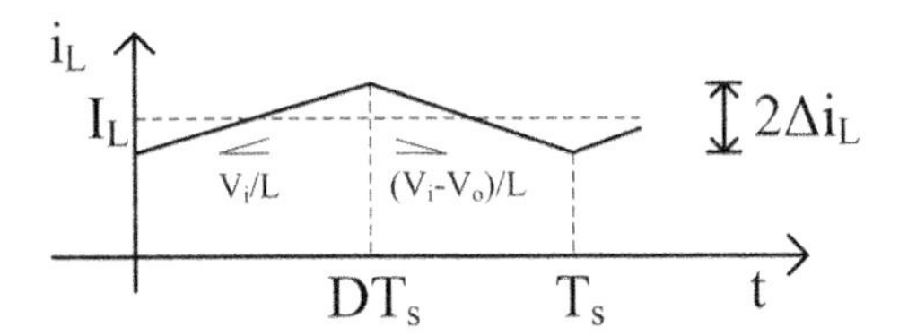

Figure 8.10 Boost converter steady-state current variation under CCM operation.

one period $D+D_1+D_2=1$. The current through the inductor in the three states is shown graphically in Figure 8.12. Comparing the inductor current curves in CCM (Figure 8.10) and DCM (Figure 8.12) modes, we can determine a boundary condition to distinguish between the two modes. In fact, the boundary between the two modes is when the average inductor current becomes equal to the current variation $I_L = \Delta i_L$. Using (8.21) and (8.23) for I_L and Δi_L respectively, we obtain

$$I_L = \Delta i_L \Rightarrow \frac{V_L}{(1-D)R_L} = \frac{V_i D T_s}{2L} \Rightarrow K = K_c = D(1-D)^2, \qquad (8.27)$$

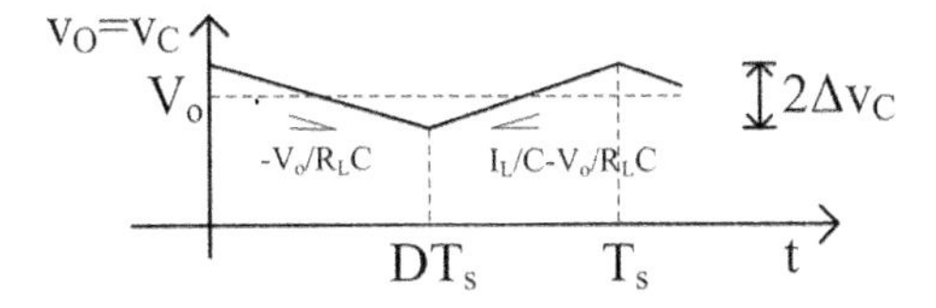

Figure 8.11 Boost converter steady-state output voltage variation under CCM operation.

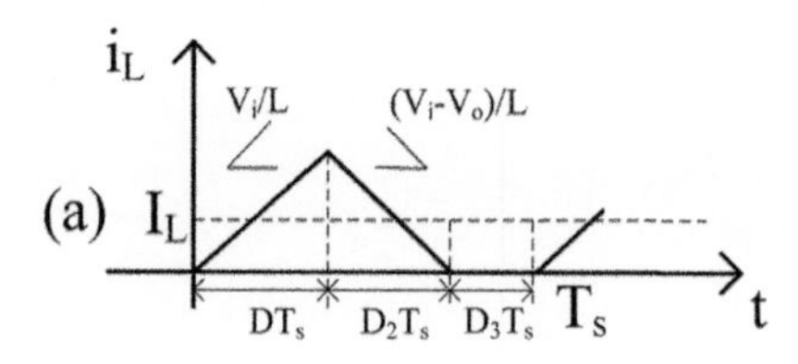

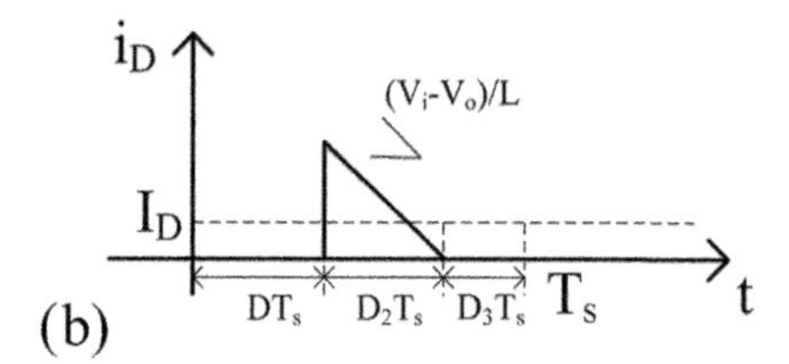

Figure 8.12 Boost converter (a) inductor and (b) diode current at DCM mode.

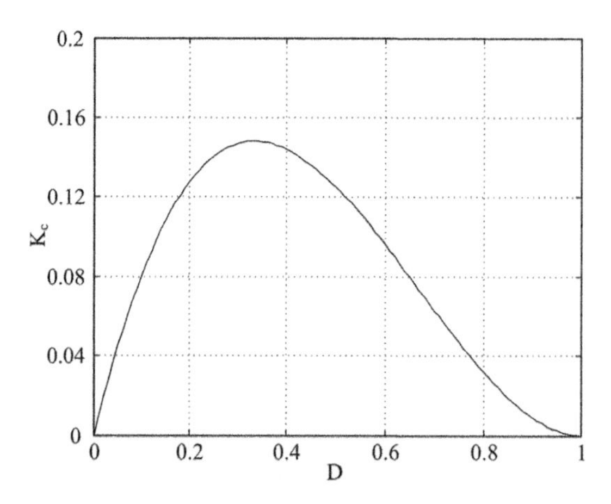

Figure 8.13 Critical value K_c defining the boundary between CCM and DCM.

where

$$K = \frac{2L}{R_L T_s}. \tag{8.28}$$

When $K > K_c$, the converter operates in CCM, and when $K < K_c$, the converter operates in DCM. The critical value $K = K_c = D(1 - D)^2$ is plotted in Figure 8.13. Selection of $K > 0.1481$ guarantees operation in CCM mode for every fraction D of the switching period.

Having defined the boundary between CCM and DCM operation, we can now analyze the DCM operation. The equivalent circuit of the boost converter corresponding to the three states is shown in Figure 8.14. We proceed in the same way as in the CCM mode by writing the inductor current and the capacitor voltage equations in the three positions. The current and voltage equations in

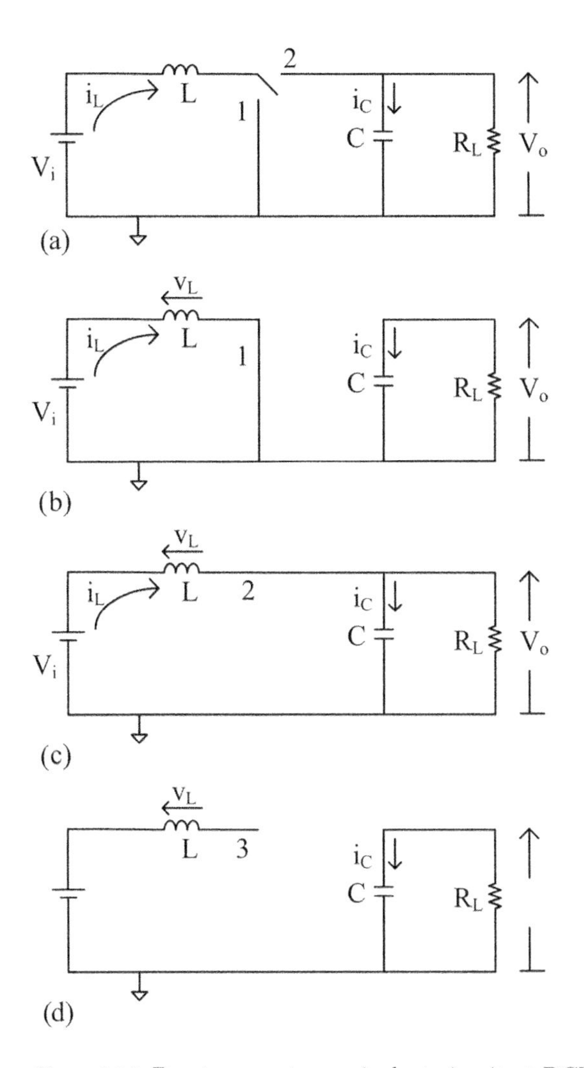

Figure 8.14 Boost converter equivalent circuit at DCM mode: (a) subinterval 1, (b) subinterval 2, and (c) subinterval 3.

the first and second switch positions at DCM are the same as in the CCM (8.17) and (8.18) valid for a fraction D and D_1 of the control signal period respectively. In the third switch position, we obtain

$$\begin{aligned} v_L &= 0 \\ i_C &= -\frac{V_L}{R_L}, \end{aligned} \tag{8.29}$$

valid for a third fraction D_2 of the control signal period. Invoking the volt-second balance for the inductor L, we have

$$\frac{1}{T_s}\int_0^{T_s} v_L(t)dt = V_i D + (V_i - V_L)D_1 + D_2 \cdot 0 = 0 \Rightarrow M(D) = \frac{D + D_1}{D_1}. \tag{8.30}$$

In this case, however, we have $D+D_1+D_2=1$, and therefore we need one more equation in order to obtain a relation between D_1 and D_2. This is done applying Kirchhoff's current law in the output node [233]:

$$i_D = i_C + \frac{v_L}{R_L} \Rightarrow < i_D >= I_D = \frac{V_L}{R_L}, \tag{8.31}$$

where we invoked the capacitor charge balance $< i_C >= 0$. The average diode current can be computed using the help of Figure 8.12. The diode current is equal to zero during position 1 and position 3, whereas it is equal to the inductor current during position 2 (Figure 8.14. The average current during position 2 is evaluated by computing the area under the triangle formed by the current plot and the time axis in Figure 8.12 and dividing it by the period T_s. The peak current I_p at the end of fraction D is found using the inductor current increase rate at position 1 from 8.26 as

$$I_p = \frac{V_i}{L} DT_s. \tag{8.32}$$

The average diode current is then found to be

$$I_d = \frac{1}{2} I_p D_1 T_s = \frac{V_i D D_1 T_s}{2L}. \tag{8.33}$$

Using this value in (8.31), we obtain

$$\frac{V_i D D_1 T_s}{2L} = \frac{V_L}{R_L} \Rightarrow V_i D D_1 = V_L K. \tag{8.34}$$

This is the equation relating D and D_1 that we needed. We can solve (8.34) for D_1 and substitute the result in (8.30) to obtain a quadratic equation for V_L

$$V_L^2 - V_L V_i - \frac{V_i^2 D^2}{K} = 0. \tag{8.35}$$

The voltage gain M_V in the DCM case is computed by solving (8.35) and selecting the positive root

$$M_V(D,K) = \frac{V_L}{V_i} = \frac{1+\sqrt{1+4D^2/K}}{2}. \tag{8.36}$$

In the event $2D/\sqrt{K} \gg 1$

$$M_V(D,K) \approx \frac{D}{\sqrt{K}}. \tag{8.37}$$

The combined voltage gain plot for CCM and DCM operation is shown in Figure 8.15.

Once we have found the voltage gain of the boost converter, we can easily compute the input resistance using (8.8) as

$$R_i = \frac{4R_L}{\left(1+\sqrt{1+4D^2/K}\right)^2}. \tag{8.38}$$

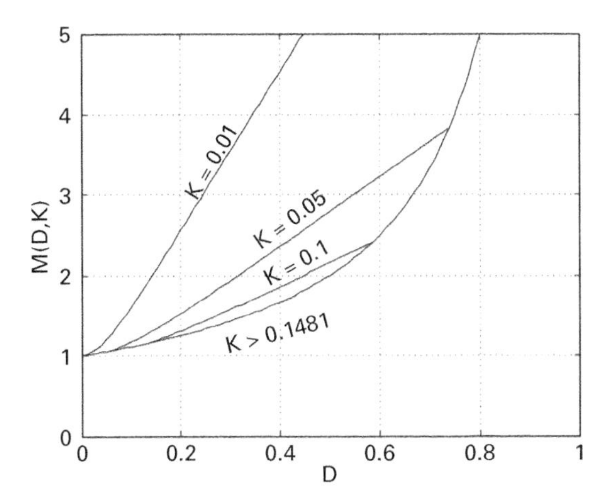

Figure 8.15 Boost converter voltage gain $M_V(D,K)$ for different values of K resulting in CCM or DCM operation.

The input resistance can be approximated as

$$R_i \approx \frac{R_L K}{D^2} \tag{8.39}$$

when $2D/\sqrt{K} \gg 1$.

8.4 Summary of Switched Mode Power Converter Properties

We can apply the small ripple approximation and volt-second and charge balance principles in the circuits of the buck and buck-boost converters in order to calculate the voltage gain and input resistance under CCM and DCM operation. The results are summarized in Tables 8.1 [233] and 8.2 [232, 233].

Table 8.1 Voltage gain of switched mode converters of Figure 8.6 [232, 233].

Converter	$K_c(D)$	$M_V(D)$ CCM	$M_V(D,K)$ DCM
Buck	$1-D$	D	$\frac{2}{1+\sqrt{1+4K/D^2}}$
Boost	$D(1-D)^2$	$\frac{1}{1-D}$	$\frac{1+\sqrt{1+4D^2/K}}{2}$
Buck-boost	$(1-D)^2$	$-\frac{D}{1-D}$	$-\frac{D}{\sqrt{K}}$
Flyback	$n^2(1-D)^2$	$\frac{D}{n(1-D)}$	$\frac{D}{n\sqrt{K}}$

In Table 8.1, n is the transformer turn ratio and $K = 2L/(T_s R_L)$ has been defined in (8.28). In the case of the flyback transformer, n is the transformer secondary to primary ratio, and the inductance $L = L_1/n^2$ used in K is the inductance L_1 of the transformer primary reflected in the secondary [232]. CCM operation corresponds to $K > K_c$.

The input resistance of the switched mode converters is found using (8.8) and summarized in Table 8.2.

Table 8.2 Input resistance of ideal switched mode converters of Figure 8.6 [232, 233, 234].

Converter	R_i CCM	R_i DCM
Buck	$\frac{1}{D^2}R_L$	$\frac{\left(1+\sqrt{1+4K/D^2}\right)^2}{4}R_L$
Boost	$(1-D)^2R_L$	$\frac{4}{\left(1+\sqrt{1+4D^2/K}\right)^2}R_L$
Buck-boost	$\frac{(1-D)^2}{D^2}R_L$	$\frac{K}{D^2}R_L$
Flyback	$\frac{(1-D)^2}{n^2D^2}R_L$	$\frac{n^2K}{D^2}R_L$

The product $KR_L = 2L/T_s$ cancels the output load R_L, and therefore the input resistance of the buck-boost and flyback converters in DCM operation is independent of the output load. It is straightforward to show that the same thing happens in buck and boost converters in DCM operation, when $2\sqrt(K)/D \gg 1$ or $2D/\sqrt{K} \gg 1$ respectively. This fact has been explored in [235] to design a 2.4 GHz rectifier circuit comprising a buck-boost converter that exhibited high efficiency over a wide range of load values.

In energy harvesting applications where the input available power is scarce, the design of the switching circuit is very challenging in order to minimize the power that is dissipated in the switching circuit itself, so that the overall circuit efficiency is not compromised. In [236], a rectifier with a boost converter operating in DCM with a very low-power switching circuit is demonstrated, harvesting ambient RF power used to power a thin film lithium battery. The boost converter operates with a $D = 0.5$ duty cycle provided by a two-stage oscillator circuit. The first oscillator is a low-frequency 250 Hz oscillator stage designed using a low-power comparator circuit. The first oscillator is used to power a second higher oscillating frequency stage with an oscillating frequency of approximately 100 KHz, which drives the gate of a power metal–oxide–semiconductor (MOS) transistor switch. The selected input resistance of the boost converter is 750 Ω. A circuit schematic representation is shown in Figure 8.16.

The oscillator circuit is powered from the output of the boost converter. One can therefore see the paradox in designing such circuits because the converter is designed to provide dc power but it also requires dc power itself in order to power the switching circuit, which is fundamental for its operation. The output energy storage component loading the converter circuit must therefore have a minimum stored energy in order for the energy harvester circuit to operate. An alternative circuit topology is to design the converter itself to comprise an oscillating input stage, which is powered by the input dc signal to the converter. Such oscillating

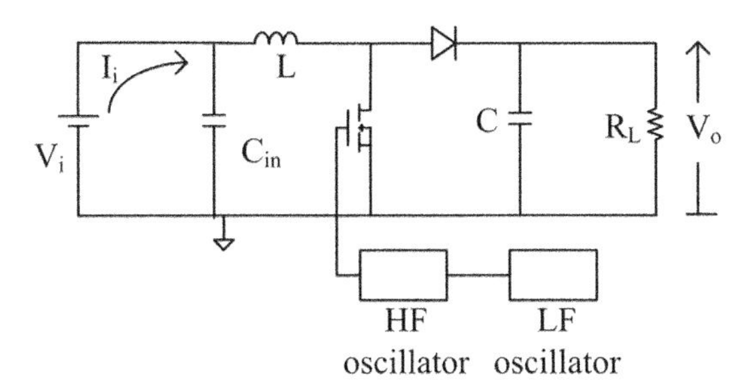

Figure 8.16 Representation of boost converter with a two-stage oscillating low-power switching circuit for ambient RF energy harvesting based on [236].

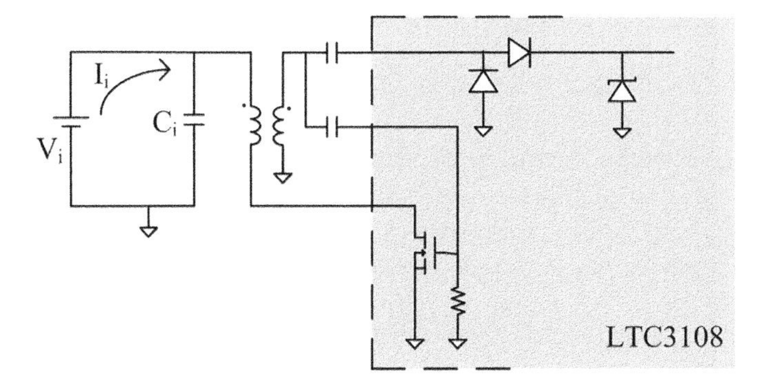

Figure 8.17 Representation of ultralow-power energy harvesting circuit LT3108 with an input oscillating converter stage, based on [106].

converter circuits have been proposed based on oscillating flyback converters that require very low dc input voltage in the order of tens of millivolts for the oscillation to start up. A popular such commercial circuit by Linear Technology designed for harvest thermoelectric energy harvesters is shown in Figure 8.17 [106]. When a transformer with a turns ratio of $n = 100$ is used, the oscillation can start with an input voltage as low as 20 mV [106]. Several similar circuits have been demonstrated in the scientific literature [237, 238, 239].

In Figure 8.17, we can also identify a Zener regulator circuit, which limits the voltage to values below 5.25 V and acts as overvoltage protection device [106]. In order to regulate the output voltage to a constant value, a feedback circuit was implemented (not shown in Figure 8.17) that sampled the output voltage using a resistive voltage divider and controlled the duty cycle of the switching waveform based on the difference between the sampled output voltage compared to a reference value. The voltage gain of the switching converter is directly dependent on the duty cycle of the switching waveform as shown in Table 8.1, and therefore the feedback circuit compensates any variations in the output voltage by adjusting the voltage gain of the converter in order to maintain a constant output voltage. The regulating control loop can be implemented either using analog circuitry or digital circuitry. In addition to regulation of the output voltage, a control loop may be designed in order to optimize the efficiency or the output power of the switching converter. Such loops are known as maximum power point tracking (MPPT) loops. MPPT tracking is commonly employed in solar energy harvesting systems [240] because the output power of the solar cell strongly depends on the output voltage, as we have seen in Figure 3.4. The tracking loop itself results in added complexity and unavoidably in added power dissipation, and therefore such loops are not typically employed in energy harvesting systems where the amount of harvested power is very low, such as in ambient RF energy harversting.

8.5 Batteries and Supercapacitors

The output of an energy harvesting is typically connected to some type of energy storage device. Commonly used energy storage devices are batteries or capacitors. Although the concept of energy harvesting is intimately related to the elimination or to the limitation of battery usage in order to mitigate the waste problems associated with spent batteries or the difficulty in certain application scenarios in substituting batteries, the time-varying nature of both the ambient energy forms and the power dissipation requirements of electronic devices requires the use of some form of energy storage device. In certain applications such as passive RFID tags [6, 241], a capacitor is sufficient to provide the required energy storage and completely eliminate the use of batteries. In typical application scenarios, however, a battery is still required. Supercapacitors share properties of capacitor and battery systems and can provide an attractive alternative to batteries. In this paragraph, we review some of the fundamental properties of battery and supercapacitor systems.

Batteries and supercapacitors are both electrochemical energy storage devices [242]. They both comprise two electrodes that are in contact with an electrolyte solution having positive and negative ions, as shown in Figure 8.18. An ion permeable separator membrane is used to separate the two electrodes in order to prevent electrical shorting between the electrodes. However, although in batteries the energy is generated by conversion of chemical energy through a redox reaction taking place at the surface of the two electrodes, this may not be the case in a supercapacitor, where ions form electrical double layers (EDLs) in the electrolyte–electrode interfaces that result in charge and consequently energy storage [242]. The two electrodes are called the anode and the cathode depending on whether they are at a higher or a lower electrical potential.

The amount of energy that is stored in an energy storage system is measured by the specific energy measured in watt-hours per kilogram (Wh/kg) or the energy density measured in watt-hours per liter (Wh/L). The power capability of the energy system is measured instead as specific power (W/kg) or power density (W/L). The different sets of energy and power measures are called gravimetric (per kg) or volumetric (per L) measures respectively. A Ragone

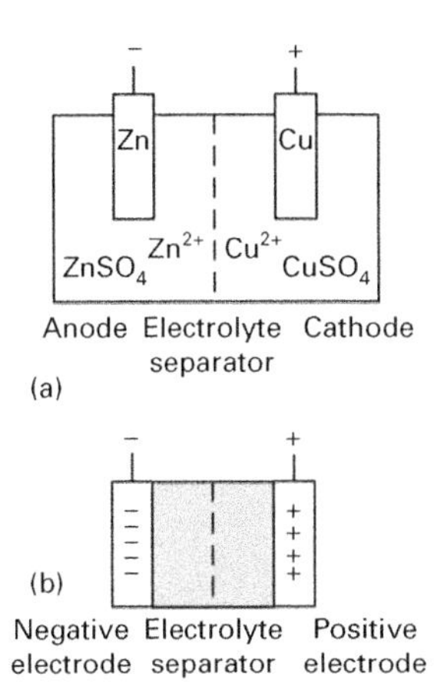

Figure 8.18 Schematic representation of (a) battery and (b) supercapacitor, based on [242].

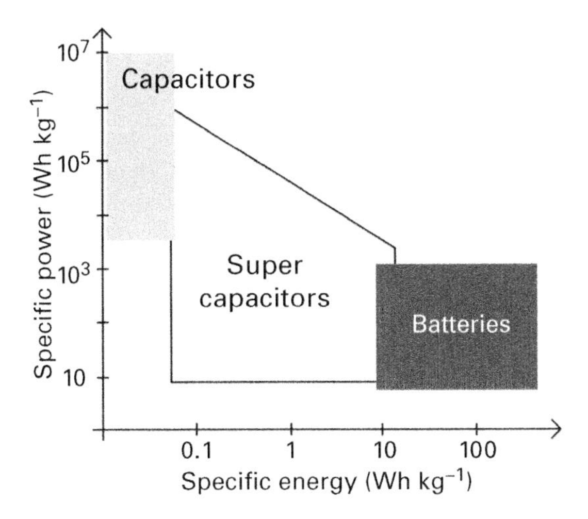

Figure 8.19 Ragone diagram of energy storage systems, based on [242].

diagram is used to compare the energy and power capability of an energy storage system. Such a Ragone diagram is shown in Figure 8.19. Batteries have a higher energy generation capability than supercapacitors, whereas supercapacitors have a higher power generation capability than batteries.

A primary battery is assembled in a charged state, and it is discharged when its terminals are connected to an electrical circuit. Primary batteries cannot be discharged. In contrast, a secondary battery is a rechargeable battery and it is usually assembled in a discharged state. Secondary batteries must be first charged before they are operated. A typical discharge curve of a battery looks like the one shown in Figure 8.20. When the terminals of a battery are open, the voltage of the battery takes its open-circuit voltage (OCV) value. When a load

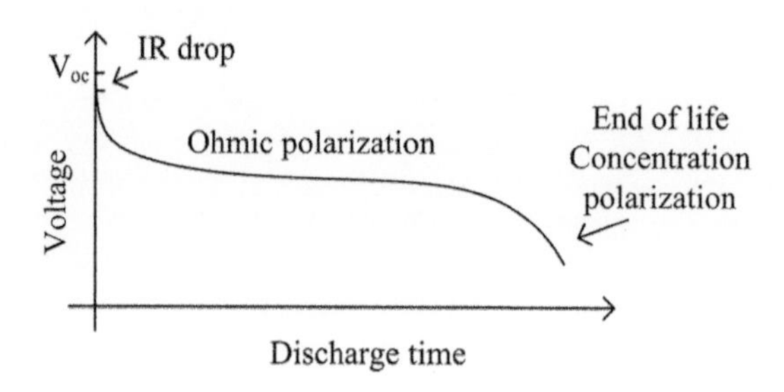

Figure 8.20 Representation of a battery discharge diagram, based on [242].

that draws some current is connected to its terminals, usually a small voltage IR drop is observed associated with the internal thermal resistance mechanisms of the battery such as electrolyte thermal resistance, contact resistance, electrode resistance, etc. The battery is then able to provide a relatively constant output voltage independent of the output current for a range of currents until a limiting value corresponding to the inability of the electrolyte redox reactions to provide further energy, where the battery output voltage begins to drop. The typical operating voltage of a battery under load is called the nominal voltage of the battery [243]. The end or cutoff voltage is the voltage of the battery at the end of the discharge. The nominal voltage of the battery depends on the battery technology and the topology (series/parallel) of the interconnected battery cells in a battery system.

The capacity C of the battery is a measure of the charge stored by the battery. It is measured in ampere-hour (Ah). Together with the nominal voltage, they represent (arguably) the most important characteristics of the battery. The C rate is typically used to represent the discharge (and the charge) current of a battery [243]. The C rate is expressed via the equation

$$I = M \times C_n, \tag{8.40}$$

where I is the current (A), C is the rated battery capacity (in Ah), n is the time base in hours for which the rated capacity is declared, and M is a multiple or fraction of C. For example, the $0.1C$ or $C/10$ discharge rate of a 1,000 mAh rated battery is 100 mA. Conversely, a 1,000 mAh rated battery discharged at the $0.2C$ or $C/5$ rate is discharged at 250 mA.

Finally, the $0.1C$ discharge rate for a battery rated at 1,000 mAh at a 5 h rate is designated as $0.1C_5 = 100$ mA [243]. Such a battery is capable of delivering 100 mA for a 5 h time interval. The capacity of the battery must not be estimated by scaling linearly a known rated value at different discharge conditions because the battery capacity typically decreases with increasing discharge current [243]. For example, the battery rated $0.1C_5 = 100$ mAh will run for more than ten hours when discharged at 50 mAh or it will run for less than one hour when discharged at its $C = 1{,}000$ mAh rate. Selected common commercial battery systems are listed in Table 8.3 [242]. It addition, lithium polymer (LiPo) batteries

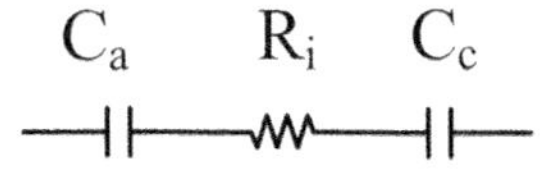

Figure 8.21 Electrical equivalent circuit of a supercapacitor.

are secondary batteries of lithium-ion technology which use a polymer electrolyte instead of a liquid electrolyte. LiPo batteries have found an increasingly large application in mobile and portable electronic devices especially due to their lower weight compared to other battery technologies.

Table 8.3 Selected commercial battery systems [242].

Name	Nominal voltage	Anode	Cathode	Electrolyte
Primary				
Alkaline	1.5	Zinc powder	Electrolytic MnO_2	aq KOH
Zinc – air	1.2	Zinc powder	Carbon (air)	aq KOH
Lithium – manganese dioxide	3.0	Lithium foil	treated MnO_2	$LiCF_3SO_3$ or $LiClO_4$
Secondary				
Lead acid	2.0	Lead	PbO_2	aq H_2SO_4
Nickel – cadmium	1.2	Cadmium	NiOOH	aq KOH
Lithium ion	4.0	Li(C)	$LiCoO_2$	$LiPF_6$

The electrical equivalent circuit of a supercapacitor, otherwise known as an ultracapacitor, is shown in Figure 8.21. The overall capacitance of the supercapacitor device is the series connection of two capacitances associated with the two double layers formed at the anode and the cathode electrode of the supercapacitor. The series resistance represents internal thermal losses of the supercapacitor, similarly to the battery. Due to the series connection of the capacitances, the overall capacitance C of a symmetric supercapacitor comprising electrodes of the same material is reduced in half compared to the capacitance of the single anode C_a and cathode C_c electrodes [242].

$$\frac{1}{C} = \frac{1}{C_a} + \frac{1}{C_c} \Rightarrow C = \frac{C_a}{2} = \frac{C_c}{2}. \tag{8.41}$$

The voltage of supercapacitors using an aqueous electrolyte is approximately 1 V, whereas using an organic-based electrolyte results in a voltage of approximately 2.7 V. Asymmetric supercapacitors comprise a battery electrode as one of the electrodes, for example the cathode electrode, which due to the redox reaction comprises a capacitance approximately ten times larger than the electrical double

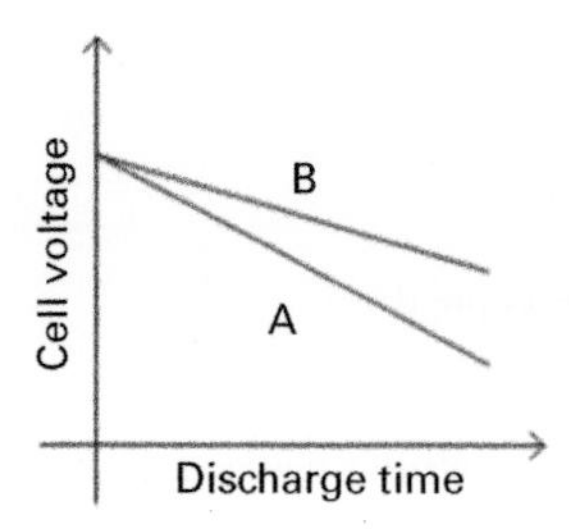

Figure 8.22 Supercapacitor discharge diagram [242].

layer capacitance. As a result, the overall capacitance of the series connection is approximately equal to the capacitance of the electrical double layer capacitance,

$$\frac{1}{C} = \frac{1}{C_a} + \frac{1}{C_c} \approx \frac{1}{C_a} + \frac{1}{10C_a} \Rightarrow C \approx C_a. \tag{8.42}$$

Asymmetric supercapacitors therefore achieve approximately double capacitance values compared to their symmetric counterparts. Finally, the output voltage of a supercapacitor reduces linearly with time, as shown in Figure 8.22. This behavior is different from batteries as compared to Figure 8.20. As we can see in Figure 8.22, the discharge slope is smaller in the case of asymmetric supercapacitors.

8.6 Problems and Questions

1. Describe what is a linear and a switch mode power converter.
2. What are the three principles used to analyze approximately the performance of switching power converters?
3. What is the inductor flux linkage balance or volt-second balance?
4. What is the capacitor charge balance or ampere-second balance?
5. Describe what is the continuous conduction mode (CCM) and the discontinuous conduction mode (DCM) of a switched mode power converter.
6. Compute the voltage gain and input resistance of a buck converter in CCM operation.
7. Compute the voltage gain and input resistance of a buck converter in DCM operation.
8. Compute the voltage gain and input resistance of a buck-boost converter in CCM operation.
9. Compute the voltage gain and input resistance of a buck-boost converter in DCM operation.
10. What is the capacity of a battery, and what is the C-rate specification of a battery?
11. How much is the $0.2C_{10}$ rate of a battery rated as 800 mAh at a 10 h rate?

9 A System Perspective

9.1 Introduction

In recent years, significant scientific and industrial efforts have been directed toward ubiquitous sensing electronics. The concepts of "smart" devices, buildings, and cities providing us with useful information about our surrounding environment in order to improve our living conditions implicitly require the installation of a very large number of sensor and actuator devices communicating wirelessly with each other and with other networking devices. The notion of such an "Internet of Things" of devices connected to the Internet has been visualized since the early 1990s [1, 2]. Energy harvesting technologies are particularly suitable for such devices, which require very low power consumption and large energy autonomy, permitting them to operate for long periods of time without (secondary) battery recharging or the need for (primary) battery substitution.

In addition to exploring harvesting the available ambient energy in its various forms, ultralow power communication, control, and sensing electronics are necessary. Integrated electronics technologies such as CMOS have already enabled the implementation of commercially available electronic circuits dissipating power in the order of a few μW and down to tens or a few hundred nanowatts. There exists, for example, a plurality nanowatt commercial comparator circuits, such as [244]. Furthermore, complete systems such as passive UHF radio frequency identification (RFID) tag integrated circuits (ICs) require as low as 6.2 μW to be read wirelessly from a reader device [245].

Having presented various energy harvesting solutions as well as power conversion circuits in the previous chapters, in this chapter we discuss the challenges for wireless sensing platforms exploring energy harvesting.

9.2 Wireless Sensing Platforms

The architecture of a wireless sensing platform with energy harvesting capability is shown in Figure 9.1 [246]. It comprises a number of transducer devices harvesting power from different energy sources such as solar, thermal, vibration, or RF and converting it to dc electrical power. A suitable dc–dc converter circuit converts the dc output voltage of the transducers to a suitable dc voltage value

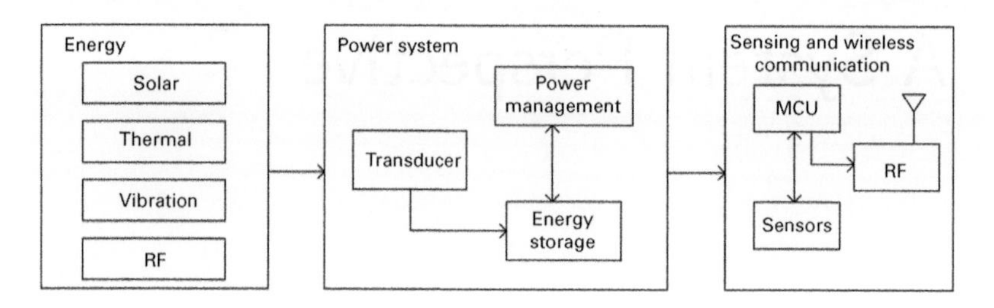

Figure 9.1 Block diagram of wireless sensing platform with energy harvesting capability.

and supplies a storage device. A power management unit (PMU) optimizes the conversion efficiency of the dc electrical power and controls the operating duty cycle of the platform. The collected power is used to power a microcontroller device that reads various sensors, computes the required information, stores to or retrieves information from memory, and finally communicates information wirelessly using a suitable transceiver device.

The combination of multiple energy harvesting transducers is preferred due to the strongly time-varying and unpredictable nature of the available ambient energy. A summary of typical values of available power density is presented in Table 9.1 [246]. Although RF ambient energy availability appears to be very low compared to other forms of energy, the easy integration of RF energy harvesters with wireless transceivers and other transducers (we have seen in previous chapters that rectennas can be easily integrated with solar cells and even thermoelectric generators) as well as the capability of integration with intentional wireless power transmission devices makes it a convenient source of energy for such platforms.

Efficient combination of dc sources that are variable and may have very different relative values is not a trivial task due to the fact that the efficiency and impedance of each source depend on the input ambient energy and also the fact that each source presents a load to the others, which can affect the combined efficiency of the system. Efficient combination of different RF energy harvesters has been addressed, for example, in [145, 223, 247]. In [247], it was demonstrated that the loading effect of the different source inputs can be mitigated by the use of additional circuit paths that include "shortcut" diode elements in order to minimize the loading effect of underperforming rectifier circuits. Efficient combination of solar cell outputs into solar modules is also discussed broadly in the literature, such as in [248, 249]. In the case of solar modules, one simple method to isolate the various cells and eliminate undesired loading is through the use of blocking diodes. In the case of low available power scenarios, one needs to carefully consider the efficiency loss associated with the power dissipated within the dc blocking or shortcut diodes. An efficient power combining circuit from a light and RF harvester was demonstrated in [250] where two different rectifier paths are included and are optimized under different light intensity and

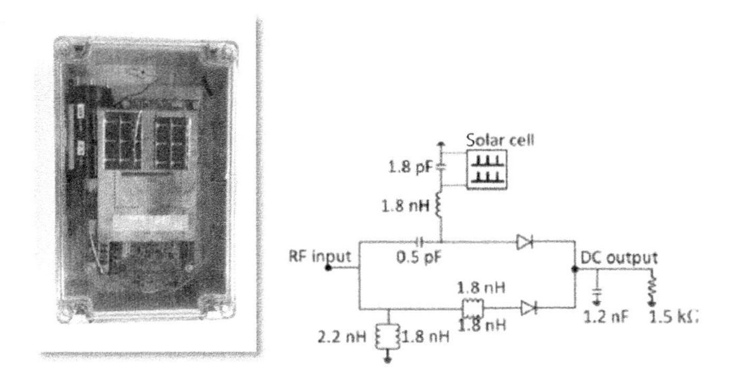

Figure 9.2 Solar and RF power combining prototype sensor [250].

RF power conditions (Figure 9.2). In general, added circuit complexity results in a potentially lower maximum achievable efficiency due to the use of more (nonideal) circuit elements, but it achieves a better average efficiency for different input energy scenarios. One further possibility when one of the two power sources is much lower than the other is the use of the low-power source as an efficiency-boosting circuit for the higher-power source, as was done by the authors in [251] by using the output of a thermoelectric energy harvester to prebias an RF energy harvesting circuit.

Table 9.1 Selected ambient energy sources and available transducers [246].

Parameter/ energy	Solar	Thermal	Vibration	RF
Power density	100 mW/cm^2	60 μW/cm^2	200 μW/cm^3	0.00002–1 μW/cm^2
Output voltage	0.9 V (a-Si cell)	20–500 mV ($\Delta T < 10$ deg)	10 V	10–100 mV
Power	60 mW	0.52 mW	8.4 mW	1.5 μW
Condition	6.3 × 3.8 cm Flex. cell AM1.5 illum.	TEG $\Delta T = 5.6°$	Piezoel. Shoe mounted	Power dens. 0.15 μW

9.3 Voltage Conversion Circuits for Energy Harvesting Transducers

The output voltage of the various energy harvesting transducer circuits varies significantly during the operating time of the transducer as well as between different transducer types as shown in Table 9.1. Therefore, dc–dc voltage conversion circuits are necessary in order to bring the output voltage to a value that is suitable for the microcontroller, sensing and wireless communication circuits of the wireless sensing platform (Table 9.1).

Table 9.2 shows some low-power voltage conversion circuits that are typically used in existing sensing platforms. The input voltage sensitivity of such circuits is typically in the order of a few hundred mV. Furthermore, typically dc–dc converter circuits present hysterisis in their behavior, which results in a higher voltage required for them to begin operation from zero input conditions, known as cold start. Once operating, the input voltage may be reduced below the cold start value maintaining operation. The LTC3107 device uses a self-oscillating converter with a transformer-based oscillator, which allows it to have a very low sensitivity and begin operation for input voltages as low as 20 mV, when a transformer with a 1:100 ratio is used [106]. Despite the very low startup voltage, the input impedance of the self-oscillating dc–dc converter is also low in the order of a few Ω [106], which makes it unsuitable for RF energy harvester comprising rectifier circuits where the optimum load impedance at low input powers is in the order of a few KΩ (see Section 7.3). In addition, it is not possible to control the input impedance, which makes it unsuitable to implement some type of maximum power point tracking (MPPT) system [252]. The output voltage is determined by the electronics of the sensing platform, and sometimes it can be changed or the converter circuit may provide a plurality of different output voltage values. The quiescent current represents the dissipated power that the converter itself requires to operate when no output load is connected to it, i.e., the microcontroller, sensor, and RF transceiver circuitry is not active.

Table 9.2 Selected low-power voltage conversion circuits for energy harvesting transducers [246].

Name	Manuf.	Quiesc. I_q	Sensitivity	V_o	EH source
LTC3107, LT3108	Analog Dev.	80 nA 200 nA	20–500 mV	2.35 V, 3.3 V 4.1 V, 5V	Thermal
BQ25505	TI	325 nA	100 mV 330 mV (cold)	5.0 V	Solar, thermal
SPV1050	STMicro		180 mV	3.6 V	Solar, thermal
MAX17710	MAXIM	625 nA	750 mV	6.0 V	RF, solar, thermal
PCC110	Powercast		−17 dBm	-	RF

9.4 Low-Power Microcontroller Units (MCU)

The TI MSP430 microcontroller family has traditionally been used for low-power applications, especially related to RFID systems [253, 254]. Other low-power microcontrollers such as the SiM3C1XX family from Silicon Laboratories or the PIC24F16KA102 from Microchip Technologies present suitable alternatives. Table 9.3 summarizes some of the available MCUs.

Table 9.3 Low-power microcontroller (MCU) circuits for wireless sensing platforms [15].

Name	Clock speed	Operating voltage	Current
TI MSP430	16 MHz 4 KHz	1.8–3.6 V	7 mA 5–6 μA
Microchip PIC24F	32 MHz 32 KHz	1.8–3.6 V	11 mA 8–15 μA
Silicon Labs SiM3C1XX	80 MHz 16.4 KHz	1.8–3.6 V	33 mA 175–250 μA

A higher clock speed results in faster computation but also in a higher dissipation. In fact, the energy E_{MCU} dissipated in an MCU that is operated for a time interval T is given by [252]

$$E_{MCU} = V_{CC}I_oT + V_{CC}m(V_{CC})TF_{CLK}, \tag{9.1}$$

where $m(V_{CC})$ is the slope of the linear curve of the current versus clock frequency F_{CLK}, which depends on the supply voltage V_{CC}. I_o is the current that is dissipated by the MCU in active mode when the clock frequency is $F_{CLK} = 0$ Hz. A typical current versus clock frequency curve is shown in Figure 9.3. The energy dissipation can be reduced by reducing the supply voltage V_{CC}, the clock frequency f_{CLK}, and the total operating time T. The computational requirements for such low-power sensing platforms are typically not heavy and therefore one can strive to use the minimum possible clock frequency. Nonetheless, there is some minimum fixed energy dissipation overhead associated with the current I_o that can be minimized by reducing as much as possible the operating interval T. This can be done by maximizing the efficiency of the implemented code programming the MCU because the number of executed instructions is proportional to Tf_{CLK} [252]. The minimum supply voltage of the MCU is limited by the voltage required for its memory, which for example is limited to 1.8 V for FRAM memory technologies.

Finally, all MCUs have active modes where they perform computation, read or write to the memory, or communicate with the peripheral devices such as sensors and wireless communication circuits and additionally power saving modes in order to be able to reduce their overall power consumption [252]. One type of

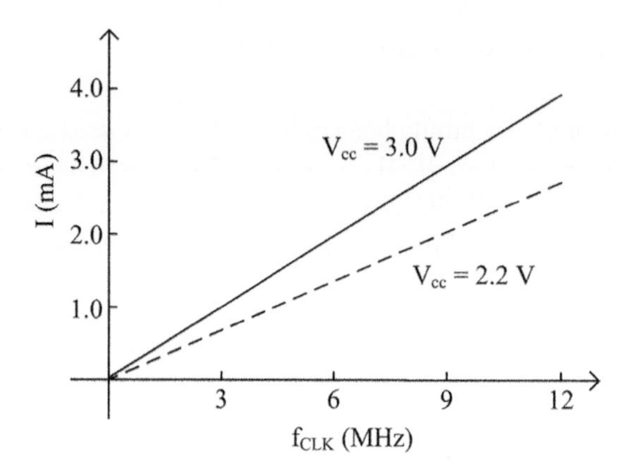

Figure 9.3 Representation of MCU current consumption versus clock frequency f_{CLK} for different supply voltage values V_{CC}, based on [252].

power saving mode is "hibernate" modes, where all clocks are stopped. A second type of power saving mode is one where only a real-time clock (RTC) oscillator with frequency in the order of 5–60 kHz operates. Hibernate modes are the least power consuming, but the MCU requires an external stimulus to wake up from this dormant state, and the energy required to operate such a sensor should be considered in the overall system energy budget. It is possible to use an external timer circuit that has a very low current consumption (e.g., 30 nA) to bring the MCU out of the hibernate mode [252]. The RTC mode is less energy efficient; however, the active oscillator requires a current in the order of 0.5 μA.

9.5 Sensor Circuits

Texas Instruments and many other companies offer a series of low-power sensors that can be used together with the various microcontroller and radio transceiver circuits for the wireless sensor platforms. One example is the digital (11 bit) low-power humidity and integrated temperature sensor HDC1000. Its average current for a humidity and temperature measurement is 1.2 μA from a 3 V supply. A variety of low-power digital or analog sensors can be considered. Nonetheless, sensor circuits may significantly increase the dissipation power requirements of the platform, and they need to be carefully selected.

Alternatively, antenna-based sensing techniques [255, 256] can be used that minimize the power dissipation. According to antenna-based sensing, an antenna can be designed such that a change in a desired sensing parameter – temperature, material permittivity, a level of some liquid, or the presence of cracks in a material, just to name a few examples – can result in a modification of an antenna parameter such as its input impedance matching or gain. The antenna is part of an RFID tag, and modification of the antenna parameter results in a reduction in the backscattered power toward the reader when the tag is interrogated, which,

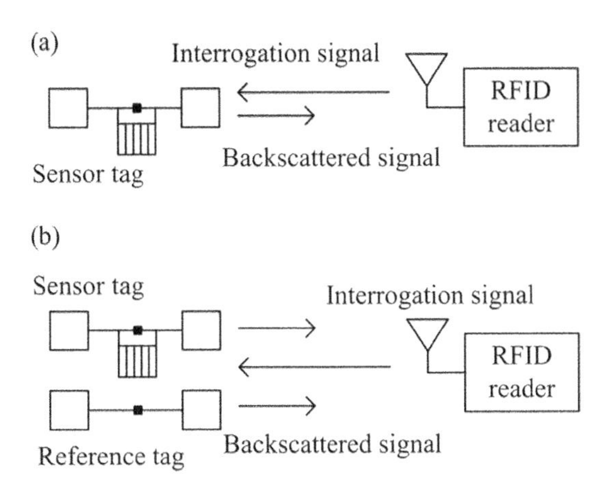

Figure 9.4 RFID tags with antenna-based proximity sensing: (a) single tag, (b) dual tag, based on [257].

in turn, can be translated by the reader in a sensing parameter. There are several challenges in an antenna-based sensing scheme regarding measurement accuracy and repeatability; however, the fact that the sensing capability is superimposed in the communication scheme without essentially requiring additional power makes it an attractive sensing option. One way to improve or facilitate the sensing measurement is to implement a differential measurement by employing two tags in proximity to each other where one tag is a reference tag and the second tag is the sensing tag. The reader interrogates both tags and determines the sensing parameter by comparing the backscattered power from the two tags. One such system implementing a proximity sensor, as shown in Figure 9.4, was demonstrated in [257].

9.6 Wireless Transceivers and Backscatter Communication

Finally, there exist also several low-power wireless communication transceiver modules that can be used in such platforms and a selected subset of them is presented in Table 9.4. Commonly used license-free frequency bands are used that range from sub-GHz range including 315 MHz, 433 MHz, and UHF RFID frequency bands of 868 MHz and 915 MHz to the 2.45 GHz industrial, scientific, and medical (ISM) band [252]. Many of the devices support the IEEE 802.15.4 standard at 868 MHz, 915 MHz, and 2.45 GHz bands, although some of the devices are designed for proprietary modulation, such as the TI CC2500 device. The transmit power is limited by the operating band regulations and typically ranges from 5 to 15 dBm, although the trade-off between power dissipation minimization and operating range maximization may further limit the transmitted power.

Table 9.4 Selected low-power transceiver circuits for wireless sensing platforms [252].

Manufacturer	Device	Frequency
Analog Devices	ADF7241	2.45 GHz
	ADF7020	433, 868, 915 MHz
Atmel	AT84RF23X	2.45 GHz
	ATA5428	433, 868 MHz
Microchip	MRF24J40	2.45 GHz
	MRF89XA	868, 915 MHz
TI	CC25XX	2.45 GHz
	CC11XX	315, 433, 868, 915 MHz
Silicon Labs	Si4420	315, 433, 868, 915 MHz

Low-power wireless communication transceivers require an amount of power that is in the order of mW when transmitting or receiving. Backscatter communication presents a very low-power alternative that does not use an active transmitter or receiver circuit, thereby minimizing the dissipated power for communication to the μW range. In backscatter radios, information is transmitted by modulating the load connected to an antenna [6]. A receiving antenna typically scatters partially an impinging wave [258]. When a load connected to the antenna is changed according to a pattern that corresponds to an information signal that we want to transmit, the signal scattered by the antenna is modulated according to the load and therefore it contains the desired information. In a passive RFID system, the signal that impinges on the antenna is an interrogating carrier signal from a reader device. The reader then receives the scattered signal from the antenna and demodulates the information that has been added to it by the tag. The tag uses the carrier signal from the reader to power itself before it can modulate the backscattered signal to the reader. A conceptual block diagram of the system is shown in Figure 9.5. Current tags require a minimum power in the order of 6.2 μW to power themselves [201]. Commercial reader devices transmit power in the order of 30 dBm depending on regulations and use antennas with approximately 7 dB of gain.

The operating range of the backscatter communication system is limited by the minimum power required to power the tag. The tag range can be estimated by applying the Friis transmission formula twice, once for the carrier signal propagating from the reader to the tag and a second time for the backscattered signal from the tag to the reader, resulting in

$$R = \frac{\lambda_o}{4\pi}\sqrt{\frac{P_t G_t G_r \tau}{P_{th}}}, \tag{9.2}$$

where λ_o is the wavelength, P_t is the transmitted power by the reader, G_t is the gain of the reader antenna, G_r is the gain of the tag antenna, P_{th} is the sensitivity of the RFID chip, and τ is the transmission coefficient [259],

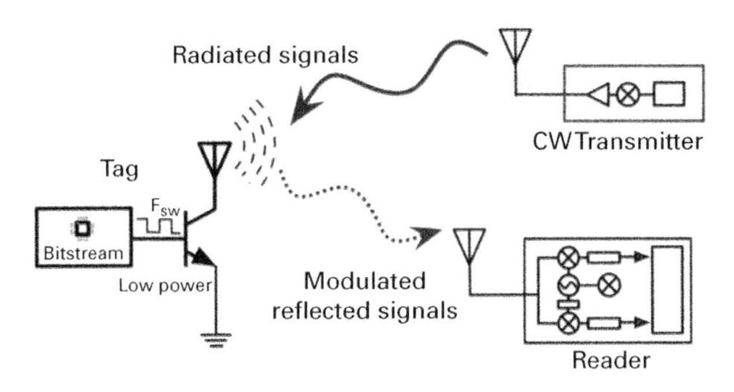

Figure 9.5 Backscatter communication.

$$\tau = \frac{4R_a R_c}{|Z_c + Z_a|^2}. \tag{9.3}$$

$Z_c = R_c + jX_c$ and $Z_A = R_A + jX_A$ are the complex impedances of the RFID chip and the tag antenna respectively. The power transmission coefficient τ reflects how good is the impedance match between the RFID tag antenna and chip and becomes equal to 1 when $Z_c = Z_A^*$, where $()^*$ denotes the complex conjugate. The range equation (9.2) can be normalized to a reference range factor R_o

$$R_o = \frac{\lambda_o}{4\pi}\sqrt{\frac{P_t G_t}{P_{th}}} \tag{9.4}$$

as

$$\frac{R}{R_o} = \sqrt{G_r \tau}. \tag{9.5}$$

Using $P_t = 30$ dBm, $P_{th} = 6\ \mu$W, $\lambda_o = 0.33$ m corresponding to an operating frequency of 915 MHz and $G_t = 7$ dB, one can compute that the reference range factor of an RFID system is $R_o = 23.8$ m. This is an optimistic estimate of the actual achievable range as typically a tag antenna may have less gain than 0 dB and also the τ parameter can be less than 1 due to an impedance mismatch. Furthermore, one should keep in mind that the propagation environment may result in additional losses that correspond to a worse performance than the one predicted by applying the Friis transmission formula.

It is possible to increase the operating range by implementing a bistatic configuration where the reader is separated in a power supply circuit that can be placed closer to the tsg and a receiver circuit that can be placed much further from the tag. This way it was possible to demonstrate experimentally an operating range between the tag and the receiver of more than 100 m [260].

Perhaps the most widely known wireless sensing platform based on backscatter communication is the wireless Internet service provider (WISP) platform. It implements a backscatter communication tag using an MSP430F1232 microcontroller, it is modular, and it can integrate a number of sensors. A block diagram of the WISP circuit is shown in Figure 9.6.

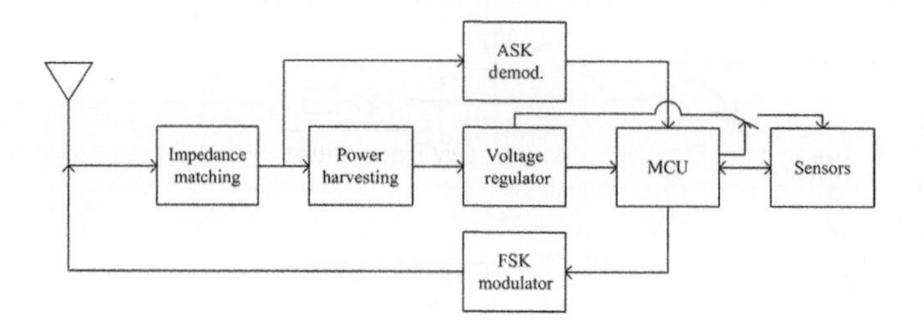

Figure 9.6 Representation of WISP circuit block diagram, based on [253].

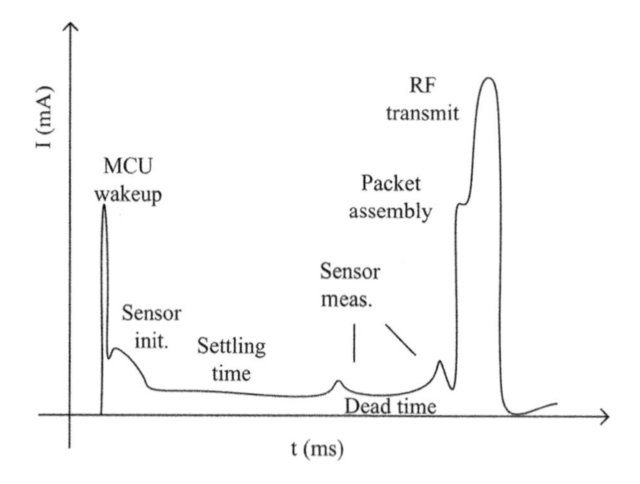

Figure 9.7 Representation of current drawn during a packet transmission, based on [252].

9.7 Energy Consumption

An example of the current drawn from a storage element during a transmission cycle of a wireless sensing platform is shown in Figure 9.7 [252]. In Figure 9.7, we can identify the different stages of the packet transmission from the wakeup of the MCU to the sensor measurements and transmission of a packet. How can we minimize the energy consumption of the system? The RF transmission power can be greatly reduced by employing a backscatter communication system if the operating range of the system permits it. A significant amount of power is dissipated in the sensor through long settling times and time intervals between measurements corresponding to a dead time. It is therefore imperative to properly select low-power sensors or if possible employ techniques such as antenna-based sensing if possible.

Assuming that the energy storage element is a capacitor with capacitance C that the duration of the interval where the system is active is T, an approximate expression of the required energy is given by [253]

$$V_{cc}(I_s + I_m)T \leq \frac{1}{2}C\left(V_{req}^2 - V_{cc}^2\right), \tag{9.6}$$

where I_s is the average current dissipated by the sensor, I_m is the average current dissipated by the MCU and the communication unit, and V_{cc} is the supply voltage. V_{req} is the required voltage that the capacitor must be charged in order to have sufficient energy to power the wireless sensor platform.

9.8 Ambient Backscattering

Backscatter communication permits us to reduce the power consumption of the wireless sensing platform by not requiring an active transceiver due to the fact that an external reader device provides the energy for the transmission and reception of information. Actually, the reader provides a carrier signal, which is used both to power the tag and to convey the information. The concept of ambient backscattering, proposed by a team of the University of Washington [261], further eliminates the need for the reader transmitting a carrier to power the tag and transfer the information back to the reader. In ambient backscattering, a tag modulates the scattered ambient signals from its antenna by changing its load. Furthermore, the tag may be powered by ambient RF signals such as FM, TV, or cellular phone signals. This is a circuit that relies on ambient RF signals to both power itself and to communicate. In [261], information bit rates of 100 bps, 1 bps, and 10 bps were used to demonstrate communication between the tag and the receiver reader for distances of a few feet in indoor and outdoor scenarios while the ambient signal was provided by a TV station.

We also implemented at Heriot-Watt University an ambient backscatter system and demonstrated communication using ambient FM signals (see Figure 9.8) [262]. The tag modulated scattered signals from an FM station in the Edinburgh area located 34.6 km away from the laboratory. The power of the FM station carrier signal was measured to be approximately −51 dBm in the vicinity of the tag antenna in the laboratory. Indoor experiments in the lab were performed, shown in Figure 9.9, where an operating distance between the tag and the receiver of a few meters was demonstrated.

9.9 Problems and Questions

1. How does the MCU current consumption depend on the supply voltage and on the clock frequency?
2. What is antenna-based sensing?
3. What are the fundamental circuit blocks of a wireless sensing platform?
4. Describe the operating principle of backscatter communication. What is the difference between a monostatic and a bistatic configuration?

Figure 9.8 Ambient backscattering experiment using an FM radio station near Glasgow located 34.6 km from Edinburgh and Heriot-Watt University. ©2017 IEEE. Reprinted with permission from [262]

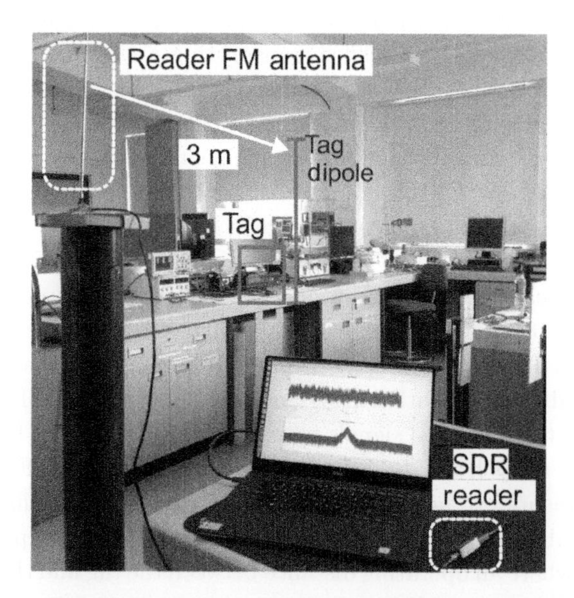

Figure 9.9 Ambient backscattering experiment lab setup at Heriot-Watt University. ©2017 IEEE. Reprinted with permission from [262]

5. What is the theoretical operating range of an 868 MHz perfectly matched RFID tag with sensitivity −20 dBm and gain −5 dB, when it is used with a reader that transmits 30 dBm of power with a 7 dB antenna (assume no polarization mismatch between the two antennas)?
6. What is ambient backscattering?
7. Given a backscatter system with an energy storage capacitor $C = 1\ \mu$F and a supply voltage of 1.8 V, what is the required minimum voltage that the capacitor must be charged to in order for it to be able to actively operate for 1 μs?

References

[1] M. Weiser, "The computer for the 21st century," *Scientific American*, vol. 265, no. 3, pp. 94–104 September 1991.

[2] R. S. Raji, "Smart networks for control," *IEEE Spectrum*, vol. 31, no. 6, pp. 49–55, June 1994.

[3] B. Warneke, M. Last, B. Liebowitz, and K. S. J. Pister, "Smart dust: communicating with a cubic-millimeter computer," *Computer*, vol. 34, no. 1, pp. 44–51, January 2001.

[4] https://en.wikipedia.org/wiki/Crossbow_Technology.

[5] J. Polastre, R. Szewczyk, and D. Culler, "Telos: enabling ultra-low power wireless research," in *Proc. 4th International Symposium on Information Processing in Sensor Networks (ISPN), Boise, ID, USA*, pp. 364–369, 2005.

[6] K. Finkenzeller, *RFID Handbook*, 3rd ed. Wiley-Blackwell, 2010.

[7] F. Boccardi, R. W. Heath, A. Lozano, T. L. Marzetta, and P. Popovski, "Five disruptive technology directions for 5G," *IEEE Communications Magazine*, vol. 52, no. 2, pp. 74–80, February 2014.

[8] A. Georgiadis, "Energy harvesting for autonomous wireless sensors and RFID's," in *Proc. XXXIth URSI General Assembly and Scientific Symposium (URSI GASS), Beijing*, pp. 1–5, 2014.

[9] Powerfilm solar cell: SP3-37 flexible solar panel, 2014.

[10] N. S. Shenk and J. A. Paradiso, "Energy scavening with shoe-mounted piezoelectronics," *IEEE Micro*, vol. 21, no. 3, pp. 30–42, May/June 2001.

[11] MicroPelt Inc., TEG MPG-D751.

[12] A. Georgiadis, G. Andia, and A. Collado, "Rectenna design and optimization using reciprocity theory and harmonic balance analysis for electromagnetic (EM) energy harvesting," *IEEE Antennas and Wireless Propagation Letters*, vol. 9, pp. 444–446, 2010.

[13] H. J. Visser, A. Reniers, and J. Theeuwes, "Ambient RF energy scavenging: GSM and WLAN power density measurements," in *Proc. European Microwave Conference, Amsterdam*, pp. 721–724, 2008.

[14] Y. Kawahara, K. Tsukada, and T. Asami, "Feasibility and potential application of power scavenging from environmental RF signals," in *Proc. IEEE AP-S Antennas and Propagation International Symposium (APSURSI), Charleston*, pp. 1–4, 2009.

[15] R. J. Vyas, B. S. Cook, Y. Kawahara, and M. M. Tentzeris, "E-WEHP: a batteryless embedded sensor-platform wirelessly powered from ambient digital-TV signals," *IEEE Transactions on Microwave Theory and Techniques*, vol. 61, no. 6, pp. 2491–2505, June 2013.

[16] S. Kim, B. Cook, T. Le, et al., "Inkjet-printed antennas, sensors and circuits on paper substrate," *IET Microwaves, Antennas & Propagation*, vol. 7, no. 10, pp. 858–868, July 2013.

[17] Y. Kawahara, S. Hodges, B. S. Cook, C. Zhang, and G. D. Abowd, "Instant inkjet circuits: lab-based inkjet printing to support rapid prototyping of UbiComp devices," in *Proc. ACM International Joint Conference on Pervasive and ubiquitous Computing (UbiComp)*, pp. 363–372, September 2013.

[18] L. Atzori, A. Iera, and G. Morabito, "The Internet of Things: a survey," *Computer Networks*, vol. 54, no. 15, pp. 2787–2805, October 2010.

[19] V. Lakafosis, A. Rida, R. Vyas, L. Yang, S. Nikolaou, and M. Tentzeris, "Progress towards the first wireless sensor networks consisting of inkjet-printed, paper-based RFID-enabled sensor tags," *Proc. IEEE*, vol. 98, no. 9, pp. 1601–1609, September 2010.

[20] M. Singh, H. Haverinen, P. Dhagat, and G. Jabbour, "Inkjet printing-process and its application," *Advanced Materials*, vol. 22, no. 6, pp. 673–685, February 2010.

[21] L. Yang, A. Rida, R. Vyas, and M. Tentzeris, "RFID tag and RF structures on a paper substrate using inkjet-printing technology," *IEEE Transactions on Microwave Theory and Techniques*, vol. 55, no. 12, pp. 2894–2901, December 2007.

[22] M. Mantysalo, P. Mansikkamaki, J. Miettinen, et al., "Evaluation of inkjet technology for electronic packaging and system integration," in *Proc. IEEE ECTC, Reno NV*, pp. 89–94, May 2007.

[23] J.-U. Park, M. Hardy, S. Kang, et al., "High-resolution electrohydrodynamic jet printing," *Nature Materials*, vol. 6, pp. 782–789, August 2007.

[24] M. Mabrook, C. Pearson, and M. Petty, "Inkjet-printed polymer films for the detection of organic vapors," *IEEE Sensors Journal*, vol. 6, no. 6, pp. 1435–1444, December 2006.

[25] K. Kordás, T. Mustonen, G. Tóth, et al., "Inkjet printing of electrically conductive patterns of carbon nanotubes," *Small*, vol. 2, no. 8–9, pp. 1021–1025, August 2006.

[26] S. Hossain, R. Luckham, A. Smith, et al., "Development of a bioactive paper sensor for detection of neurotoxins using piezoelectric inkjet printing of sol-gel-derived bioinks," *Anal. Chem.*, vol. 81, pp. 5474–5483, June 2009.

[27] Y. Liu, K. Varahramyan, and T. Cui, "Lowvoltage allpolymer fieldeffect transistor fabricated using an inkjet printing technique," *Macromolecular Rapid Communications*, vol. 26, no. 26, pp. 1955–1959, December 2005.

[28] S. Bidoki, J. Nouri, and A. Heidari, "Inkjet deposited circuit components," *Journal on Micromechanics and Microengineering*, vol. 20, no. 5, p. 055023, 2010.

[29] B. Mahar, C. Laslau, R. Yip, and Y. Sun, "Development of carbon nanotube-based sensors – a review," *IEEE Sensors Journal*, vol. 7, no. 3, pp. 266–284, February 2007.

[30] E. Hill, A. Vijayaragahvan, and K. Novoselov, "Graphene sensors," *IEEE Sensors Journal*, vol. 11, no. 12, pp. 3161–3170, December 2011.

[31] X. Wang and J. Shi, *Piezoelectric Nanogenerators for Self-Powered Nanodevices*, G. Ciofani and A. Menciassi, Eds. Springer Berlin Heidelberg, 2012.

[32] U. Caglar, K. Kimmo, and P. Mansikkamaki, "Analysis of mechanical performance of silver inkjet-printed structures," in *Proc. IEEE INEC, Shanghai, China*, pp. 851–856, 2008.

[33] K. Maekawa, K. Yamasaki, T. Niizeki, et al., "Drop-on-demand laser sintering with silver nanoparticles for electronics packaging," *IEEE Transactions on Components, Packaging and Manufacturing Technology*, vol. 2, no. 5, pp. 868–877, May 2012.

[34] B. Polzinger, F. Schoen, V. Matic, et al., "UV-sintering of inkjet-printed conductive silver tracks," in *Proc. IEEE NANO, Portland, OR, USA*, pp. 201–204, 2011.

[35] M. Allen, A. Alastalo, M. Suhonen, T. Mattila, J. Leppäniemi, and H. Seppa, "Contactless electrical sintering of silver nanoparticles on flexible substrates," *IEEE Transactions on Microwave Theory and Techniques*, vol. 59, no. 5, pp. 1419–1429, May 2011.

[36] B. Cook and A. Shamim, "Inkjet printing of novel wideband and high gain antennas on low-cost paper substrate," *IEEE Transactions on Antennas and Propagation*, vol. 60, no. 9, pp. 4148–4156, September 2012.

[37] S. Kim, C. Mariotti, F. Alimenti, et al., "No battery required: perpetual RFID-enabled wireless sensors for cognitive intelligence applications," *IEEE Microwave Magazine*, vol. 14, no. 5, pp. 66–77, July–August 2013.

[38] J. Perelaer, A. W. M. de Laat, C. E. Hendriks, and U. S. Schubert, "Inkjet-printed silver tracks: low temperature curing and thermal stability investigation," *Journal of Materials Chemistry*, vol. 18, pp. 3209–3215, 2008.

[39] Y. Li, R. Torah, S. Beeby, and J. Tudor, "An all-inkjet printed flexible capacitor for wearable applications," in *Proc. Symposium on Design, Test, Integration and Packaging of MEMS/MOEMS (DTIP), Cannes, FR*, April 2012, pp. 192–195.

[40] S. Ko, J. Chung, H. Pan, C. Grigoropoulos, and D. Poulikakos, "Fabrication of multilayer passive and active electric components on polymer using inkjet printing and low temperature laser processing," *Sensors and Actuators A: Physical*, vol. 134, no. 1, pp. 161–168, February 2007.

[41] B. J. Kang, C. K. Lee, and J. H. Oh, "All-inkjet-printed electrical components and circuit fabrication on a plastic substrate," *Microelectronic Engineering*, vol. 97, pp. 251–254, September 2012.

[42] B. Cook, J. Cooper, and M. Tentzeris, "Multi-layer RF capacitors on flexible substrates utilizing inkjet printed dielectric polymers," *IEEE Microwave and Wireless Components Letters*, vol. 23, no. 7, pp. 353–355, July 2013.

[43] www.microchem.com/pdf/SU-8-table-of-properties.pdf.

[44] Y. Lee, Y. Jang, K. Cho, S. Pyo, D.-H. Hwang, and S. Hong, "Preparation of poly(4-hydroxystyrene) based functional block copolymer through living radical polymerization and its nanocomposite with $BaTiO_3$ for dielectric material," *Journal of Nanoscience and Nanotechnology*, vol. 9, no. 12, pp. 7161–7166, December 2009.

[45] B.-J. Gans, L. Xue, U. Agarwal, and U. Schubert, "Ink-jet printing of linear and star polymers," *Macromolecular Rapid Communications*, vol. 26, no. 4, pp. 310–314, February 2005.

[46] S. Keller, G. Blagoi, M. Lillemose, D. Haefliger, and A. Boisen, "Processing of thin SU-8 films," *Journal on Micromechanics and Microengineering*, vol. 18, no. 12, pp. 1–10, November 2008.

[47] K.-J. Baeg, D. Kim, J. Kim, et al., "Flexible complementary logic gates using inkjet-printed polymer field-effect transistors," *IEEE Electron Device Letters*, vol. 34, no. 1, pp. 126–128, January 2013.

[48] E. Tekin, B.-J. Gans, and U. Schubert, "Ink-jet printing of polymers – from single dots to thin film libraries," *Journal of Materials Chemistry*, vol. 14, pp. 2627–2632, 2004.

[49] P. Yunker, T. Still, M. Lohr, and A. Yodh, "Suppression of the coffee-ring effect by shape-dependent capillary interactions," *Nature*, vol. 476, pp. 308–311, August 2011.

[50] Y. Wang and J. Yeow, "A review of carbon nanotubes-based gas sensors," *Journal of Sensors*, vol. 2009, pp. 1–24, May 2009.

[51] T. Le, V. Lakafosis, Z. Lin, and C. Wong, "Inkjet-printed graphene-based wireless gas sensor modules," in *Proc. IEEE ECTC, San Diego, CA, USA*, pp. 1003–1008, 2012.

[52] T. Le, V. Lakafosis, S. Kim, et al., "A novel graphene-based inkjet-printed WISP-enabled wireless gas sensor," in *Proc. European Microwave Conference (EuMC), Amsterdam, Netherlands*, pp. 412–415, October 2012.

[53] L. Yang, R. Zhang, D. Staiculescu, C. Wong, and M. Tentzeris, "A novel conformal RFID-enabled module utilizing inkjet-printed antennas and carbon nanotubes for gas-detection applications," *IEEE Antennas and Wireless Propagation Letters*, vol. 8, pp. 653–656, 2009.

[54] R. Martel, T. Schmidt, H. Shea, T. Hertel, and P. Avouris, "Single- and multi-wall carbon nanotube field-effect transistors," *Applied Physics Letters*, vol. 73, pp. 2447–2449, October 1998.

[55] M. Najari, S. Frégonése, C. Maneux, H. Mnif, N. Masmoudi, and T. Zimmer, "Schottky barrier carbon nanotube transistor: compact modeling, scaling study, and circuit design applications," *IEEE Transactions on Electron Devices*, vol. 58, no. 1, pp. 195–205, January 2011.
[56] A. Das, S. Pisna, B. Chakraborty, et al., "Monitoring dopants by raman scattering in an electrochemically top-gated graphene transistor," *Nature Nanotechnology*, vol. 3, pp. 210–215, March 2008.
[57] A. Manbachi and R. Cobbold, "Development and application of piezoelectric materials for ultrasound generation and detection," *Ultrasound*, vol. 19, no. 4, pp. 187–196, 2011.
[58] Z. Wang and J. Song, "Piezoelectric nanogenerators based on Zinc oxide nanowire arrays," *Science*, vol. 312, pp. 242–246, April 2006.
[59] R. Yang, Y. Qin, L. Dai, and Z. Wang, "Power generation with laterally packaged piezoelctric fine wires," *Nature Nanotechnology*, vol. 4, pp. 34–39, November 2008.
[60] G. Zhu, A. Wang, Y. Liu, Y. Zhou, and Z. Wang, "Functional electrical stimulation by nanogenerator with 58V output voltage," *Nano Letters*, vol. 12, pp. 3086–3090, May 2012.
[61] J. Nelson, *The Physics of Solar Cells.* Imperial College Press, 2003.
[62] E. Bequerel, "Memoire sur les effets electriques produits sous l'influence des rayons solaires," *Compte Rendu des Seances de l'Academie des Sciences*, pp. 561–567, Novembre 1839.
[63] S. Beeby and N. White, *Energy Harvesting for Autonomous Systems.* Artech House, 2010.
[64] S. Sze and K. Ng, *Physics of Semiconductor Devices*, 3rd ed. Wiley, 2007.
[65] M. Green, *Third Generation Photovoltaics, Advanced Solar Energy Conversion.* Springer-Verlag, 2006.
[66] E. H. Sargent, "Colloidal quantum dot solar cells," *Nature Photonics*, vol. 6, pp. 133–135, 2012.
[67] G. Konstantatos and E. H. Sargent, "Solution-processed quantum dot photodetectors," *Proceedings of the IEEE*, vol. 97, no. 10, pp. 1666–1683, October 2009.
[68] E. L. A. Vitruani and M. Powalla, "Influence of the light source on the low-irradiance performance of Cu(In,Ga)Se_2 solar cells," *Solar Energy Materials and Solar Cells*, vol. 90, pp. 2141–2149, 2006.
[69] C. Gueymard, "The sun's total and spectral irradiance for solar energy applications and solar irradiation models," *Solar Energy*, vol. 76, no. 4, pp. 423–453, 2004.
[70] www.nrel.gov/grid/solar-resource/spectra.html
[71] G. P. M. Pagliaro and R. Cirimina, *Flexible Solar Cells.* Wiley-VCH, 2006.
[72] W. Shockley and H. J. Queisser, "Detailed balancd limit of efficiency of p-n junction solar cells," *Journal of Applied Physics*, vol. 32, no. 3, pp. 510–519, 1961.

[73] C. H. Henry, "Limiting efficiencies of ideal single and multiple energy gap terrestrial solar cells," *Journal of Applied Physics*, vol. 51, pp. 4494–4500, 1980.

[74] M. A. Green, Y. Hishikawa, E. D. Dunlop, D. H. Levi, J. Hohl-Ebinger, M. Yoshita, and A. W. Ho-Baillie, "Solar cell efficiency tables (version 53)," *Progress in Photovoltaics*, vol. 27, pp. 3–12, 2019.

[75] A. de Vos, "Detailed balance limit of the efficiency of tandem solar cells," *Journal of Physics D: Applied Physics*, vol. 13, no. 5, pp. 839–846, 1980.

[76] A. S. Brown and M. A. Green, "Detailed balance limit for the series constrained two terminal tandem solar cell," *Physica E*, vol. 14, pp. 96–100, 2002.

[77] M. Tanaka, R. Suzuki, Y. Suzuki, and K. Araki, "Microstrip antenna with solar cells for microsatellites," in *IEEE Antennas and Propagation Society International Symposium AP-S*, vol. 2, June 1994, pp. 786–789.

[78] M. Zawadzki and J. Huang, "Integrated rf antenna and solar array for spacecraft application," in *IEEE Phased Array Systems and Technology*, May 2000, pp. 239–242.

[79] S. Vaccaro, J. Mosig, and P. de Maagt, "Two advanced solar antenna 'SOLANT' designs for satellite and terrestrial communications," *IEEE Transactions on Antennas and Propagation*, vol. 51, no. 8, pp. 2028–2034, 2003.

[80] M. Danesh and J. Long, "Compact solar cell ultra-wideband dipole antenna," in *IEEE Antennas and Propagation Society International Symposium AP-S*, pp. 1–4, 2010.

[81] M. Roo-Ons, S. Shynu, M. Ammann, S. McCormack, and B. Norton, "Transparent patch antenna on a-Si thin film glass solar module," *Electronics Letters*, vol. 47, no. 2, pp. 85–86, January 2011.

[82] A. Georgiadis, A. Collado, S. Via, and C. Meneses, "Flexible hybrid solar/EM energy harvester for autonomous sensors," in *IEEE MTT-S International Microwave Symposium Digest*, pp. 1–4, June 2011.

[83] F. Declercq, A. Georgiadis, and H. Rogier, "Wearable aperture coupled shorted solar antenna for remote tracking and monitoring applications," in *5th European Conference on Antennas and Propagation (EUCAP)*, pp. 1–4, April 2011.

[84] K. Niotaki, A. Collado, A. Georgiadis, S. Kim, and M. M. Tentzeris, "Solar/electromagnetic energy harvesting and wireless power transmission," *Proceedings of the IEEE*, vol. 102, no. 11, pp. 1712–1722, November 2014.

[85] A. Collado and A. Georgiadis, "Conformal hybrid solar and electromagnetic (EM) energy harvesting rectenna," *IEEE Transactions on Circuits and Systems I: Regular Papers*, vol. 60, no. 8, pp. 2225–2234, August 2013.

[86] T. Wu, R. Li, and M. Tentzeris, "A scalable solar antenna for autonomous integrated wireless sensor nodes," *IEEE Antennas and Wireless Propagation Letters*, vol. 10, pp. 510–513, 2011.

[87] A. P. Sample, J. Braun, A. Parks, and J. R. Smith, "Photovoltaic enhanced UHF RFID tag antennas for dual purpose energy harvesting," in *Proc. IEEE International Conference on RFID, Orlando*, pp. 146–153, 2011.

[88] A. Georgiadis and A. Collado, "Improving range of passive RFID tags utilizing energy harvesting and high efficiency class-E oscillators," in *Proc. 6th European Conference on Antennas and Propagation (EuCAP), Prague*, pp. 3455–3458, 2012.

[89] T. Starner, "Human-powered wearable computing," *IBM Systems Journal*, vol. 35, no. 3 and 4, pp. 1–12, 1996.

[90] S. Meninger, J. Mur-Miranda, R. Ammirtharajah, A. Chandrakasan, and L. Lang, "Vibration-to-electric energy conversion," *IEEE Transactions on Very Large Scale Integration (VLSI) Systems*, vol. 9, no. 1, pp. 64–76, 2001.

[91] S. Roundy, "Energy scavenging for wireless sensor nodes with a focus on vibration to electricity conversion," Ph.D. dissertation, University of California, Berkeley, 2003.

[92] H. Losty and D. Lewis, "A discussion on recent advances in heavy electrical plant," *Philosophical Transactions of the Royal Society of London, Series A, Mathematical and Physical Sciences*, vol. 275, no. 1248, pp. 69–75, 1973.

[93] N. Tesla, "Dynamo electric machine," US Patent US359748A, 1887.

[94] R. Amirtharajah and A. Chandrakasan, "Self-powered signal processing using vibration-based power generation," *IEEE Journal of Solid-State Circuits*, vol. 33, no. 5, pp. 687–695, May 1998.

[95] H. Tzou, *Piezoelectric Shells, Distributed Sensing and Control of Continua*. Kluwer Academic Publishers, 1993.

[96] S. Priya, H.-C. Song, Y. Zhou, et al., "A review on piezoelectric energy harvesting: materials, methods and circuits," *Energy Harvesting and Systems*, vol. 4, no. 1, pp. 3–39, 2017.

[97] C. Williams and R. Yates, "Analysis of a micro-electric generator for microsystems," *Sensors and Actuators A*, vol. 52, no. 1–3, pp. 8–11, 1996.

[98] S. Roundy, E. Leland, J. Baker, et al., "Improving power output for vibration-based energy scavengers," *IEEE Pervasive Computing*, vol. 4, no. 1, pp. 28–36, January–March 2005.

[99] S. Beeby, M. Tudor, and N. White, "Energy harvesting vibration sources for microsystems applications," *Measurement Science and Technology*, vol. 17, pp. R175–R195, 2006.

[100] S. Roundy, "On the effectiveness of vibration energy harvesting," *Journal of Intelligent Material Systems and Structures*, vol. 16, pp. 809–823, October 2005.

[101] S. Roundy and P. K. Wright, "A piezoelectric vibration based generator for wireless electronics," *Smart Materials and Structures*, vol. 13, pp. 1131–1142, 2004.

[102] L. Y. G. Orecchini, M. M. Tentzeris and L. Roselli, "'Smart shoe': an autonomous inkjet-printed RFID system scavenging walking energy,"

in *2011 IEEE International Symposium on Antennas and Propagation (APSURSI)*, Spokane WA, USA, pp. 1417–1420, 2011.

[103] The Seiko AGS Quartz Watch, https://global.epson.com/company/corporate_history/milestone_products/19_ags.html.

[104] S. Beeby, M. J. Tudor, E. Koukharenko, et al., "Micromachined silicon generator for harvesting power from vibrations," in *Proc. Transducers*, pp. 780–783, 2005.

[105] X. Gu, S. Hemour, and K. Wu, "Integrated cooperative radiofrequency (RF) and kinetic energy harvester," in *Proc. IEEE Wireless Power Transfer Conference (WPTC)*, pp. 1–3, 2017.

[106] D. Salerno, "Ultralow voltage energy harvester uses thermoelectric generator for battery free wireless sensors," *Linear Technology Technical Journal of Analog Innovation*, vol. 20, no. 3, pp. 1–10, October 2010.

[107] F. P. Incropera, D. P. DeWitt, T. L. Bergman, and A. S. Lavine, *Fundamentals of Heat and Mass Transfer*, 6th ed. Wiley, 2010.

[108] D. M. Rowe, Ed., *Handbook of Thermoelectrics.* CRC Press, 1995.

[109] S. Lineykin and S. Ben-Yaakov, "Modeling and analysis of thermoelectric modules," *IEEE Transactions on Industry Applications*, vol. 43, no. 2, pp. 505–512, 2007.

[110] J. A. Chavez, J. A. Ortega, J. Salazar, A. Turo, and M. J. Garcia, "SPICE model of thermoelectric elements including thermal effects," in *Proc. 17th IEEE Instrumentation and Measurement Technology Conference, Baltimore, MD, USA*, vol. 2, pp. 1019–1023, May 2000.

[111] S. Kotanagi, A. Matoge, Y. Yoshida, F. Utsunomiya, and M. Kishi, "Watch provided with thermoelectric generation unit," Patent No. WO/1999/019775, 1999.

[112] K. Niotaki, A. Georgiadis, and A. Collado, "Thermal energy harvesting for power amplifiers," in *Proc. IEEE Radio and Wireless Symposium (RWS), Austin, TX*, pp. 196–198, 2013.

[113] M. Virili, A. Georgiadis, K. Niotaki, et al., "Design and optimization of an antenna with thermo-electric generator (TEG) for autonomous wireless nodes," in *Proc. IEEE RFID Technology and Applications Conference (RFID-TA), Tampere*, pp. 21–25, 2014.

[114] M. Virili, A. Georgiadis, A. Collado, P. Mezzanotte, and L. Roselli, "EM characterization of a patch antenna with thermo-electric generator and solar cell for hybrid energy harvesting," in *Proc. IEEE Radio and Wireless Symposium (RWS), San Diego,* pp. 44–46, 2015.

[115] S. Lemey, "Textile antennas as hybrid energy-harvesting platforms," *Proceedings of the IEEE*, vol. 102, no. 11, pp. 1833–1857, November 2014.

[116] J. Ebert and M. Kazimierczuk, "Class E high-efficiency tuned power oscillator," *IEEE Journal of Solid-State Circuits*, vol. 16, no. 2, pp. 262–266, April 1981.

[117] T. Mader and Z. Popovic, "The transmission-line high-efficiency class-E amplifier," *IEEE Microwave and Guided Wave Letters*, vol. 5, no. 9, pp. 290–292, September 1995.

[118] Z. Popovic and J. Garcia, "Microwave class-e power amplifiers: A brief review of essential concepts in high-frequency class-E PAs and related circuits," *IEEE Microwave Magazine*, vol. 19, no. 5, pp. 54–66, July–August 2018.

[119] W. Brown, "Status of the microwave power transmission components for the solar power satellite," *IEEE Transactions on Microwave Theory and Techniques*, vol. 29, no. 12, pp. 1319–1327, 1981.

[120] C. A. Balanis, *Antenna Theory: Analysis and Design*, 4th ed. Wiley, 2016.

[121] W. C. Brown and E. E. Eves, "Beamed microwave power transmission and its application to space," *IEEE Transactions on Microwave Theory and Techniques*, vol. 40, no. 6, pp. 1239–1250, June 1992.

[122] N. Shinohara, K. Nishikawa, T. Seki, and K. Hiraga, "Development of 24 GHz rectennas for fixed wireless access," in *Proc XXXth URSI General Assembly and Scientific Symposium (GASS)*, pp. 1–4, August 2011.

[123] R. McCormmach, "H. hertz," in *Dictionary of Scientific Biography*, C. C. Gillispie, Ed., vol. VI. Charles Scribner's Sons, pp. 340–349, 1970–1980.

[124] N. Tesla, "The transmission of electric energy without wires," *Scientific American Supplement*, pp. 23, 760–23, 761, June 1904.

[125] ——, *Experiments with Alternate Current of High Potential and High Frequency*. W. J. Johnston Company Ltd, 1892.

[126] W. Brown, "The history of power transmission by radio waves," *IEEE Transactions on Microwave Theory and Techniques*, vol. 32, no. 9, pp. 1230–1242, 1984.

[127] W. C. Brown, R. H. George, and N. I. Heeman, "Microwave to dc converter," US Patent US3434678, March 1969.

[128] P. Glaser, "Power from the sun," *Science*, no. 162, pp. 857–886, 1968.

[129] H. Matsumoto, "Research on solar power satellites and microwave power transmissions in Japan," *IEEE Microwave Magazine*, pp. 36–45, 2002.

[130] J. O. McSpadden and J. Mankins, "Space solar power programs and microwave wireless power transmission technology," *IEEE Microwave Magazine*, vol. 3, no. 4, pp. 46–57, December 2002.

[131] A. Collado and A. Georgiadis, "Conformal hybrid solar and electromagnetic (EM) energy harvesting rectenna," *IEEE Transactions on Circuits and Systems I: Regular Papers*, vol. 60, no. 8, pp. 2225–2234, August 2013.

[132] J. A. Hagerty, F. Helmbrecht, W. McCalpin, R. Zane, and Z. Popovic, "Recycling ambient microwave energy with broadband rectenna arrays," *IEEE Transactions on Microwave Theory and Techniques*, vol. 52, no. 3, pp. 1014–1024, March 2004.

[133] E. Falkenstein, D. Costinett, R. Zane, and Z. Popovic, "Far-field RF-powered variable duty cycle wireless sensor platform," *IEEE Transactions*

on *Circuits and Systems II: Express Briefs*, vol. 58, no. 12, pp. 822–826, December 2011.
[134] T. Paing, J. Morroni, A. Dolgov, et al., "Wirelessly powered wireless sensor platform," in *Proc. European Microwave Conference*, pp. 999–1002, 2007.
[135] V. Rizzoli, D. Masotti, N. Arbizzani, and A. Costanzo, "CAD procedure for predicting the energy received by wireless scavenging systems in the near- and far- field regions," in *Proc. IEEE MTT-S International Microwave Symposium (IMS)*, pp. 1768–1771, May 2010.
[136] C.-H. Ahn, S.-W. Oh, and K. Chang, "A high gain rectifying antenna combined with reflectarray for 8 GHz wireless power transmission," in *Proc. IEEE AP-S Antennas and Propagation International Symposium (APSURSI)*, pp. 1–4, June 2009.
[137] J. McSpadden, L. Fan, and K. Chang, "Design and experiments of a high conversion efficiency 5.8 GHz rectenna," *IEEE Transactions on Microwave Theory and Techniques*, vol. 46, no. 12, pp. 2053–2060, December 1998.
[138] B. Strassner and K. Chang, "A circularly polarized rectifying antenna array for wireless microwave power transmission with over 78 percent efficiency," *Proc. IEEE MTT-S International Microwave Symposium (IMS)*, vol. 3, pp. 1535–1538, 2002.
[139] K. Fujimori, S. Tamaru, K. Tsuruta, and S. Nogi, "The influences of diode parameters on conversion efficiency of RF-dc conversion circuit for wireless power transmission system," in *Proc. 41st European Microwave Conference*, pp. 57–60, October 2011.
[140] Y.-J. Ren and K. Chang, "New 5.8 GHz circularly polarized retro-directive rectenna arrays for wireless power transmission," *IEEE Transactions on Microwave Theory and Techniques*, vol. 54, no. 7, pp. 2970–2976, July 2006.
[141] T.-W. Yoo and K. Chang, "Theoretical and experimental development of 10 and 35 GHz rectennas," *IEEE Transactions on Microwave Theory and Techniques*, vol. 40, no. 6, pp. 1259–1266, June 1992.
[142] K. Hatano, N. Shinohara, T. Mitani, K. Nishikawa, T. Seki, and K. Hiraga, "Development of class-F load rectennas," in *Proc. IEEE MTT-S Internstional Microwave Workshop Series on Innovative Wireless Power Transmission: Technologies, Systems and Applications (IMWS)*, pp. 251–254, May 2011.
[143] N. Shinohara and H. Matsumoto, "Experimental study of large rectenna array for microwave energy transmission," *IEEE Transactions on Microwave Theory and Techniques*, vol. 46, no. 3, pp. 261–268, March 1998.
[144] ——, "Dependence of dc output of a rectenna array on the method of interconnection of its array elements," *Wiley Electrical Engineering in Japan*, vol. 125, no. 1, pp. 9–17, October 1998.
[145] Z. Popovic, S. Korhummel, S. Dunbar, et al., "Scalable RF energy harvesting," *IEEE Transactions on Microwave Theory and Techniques*, vol. 62, no. 4, pp. 1046–1056, April 2014.

[146] T. Takahashi, T. Mizuno, M. Sawa, T. Sasaki, T. Takahashi, and N. Shinohara, "Development of phased array for high accurate microwave power transmission," in *Proc. IEEE MTT-S Internstional Microwave Workshop Series on Innovative Wireless Power Transmission: Technologies, Systems and Applications (IMWS)*, pp. 157–160, May 2011.

[147] M. S. Trotter, J. D. Griffin, and G. D. Durgin, "Power-optimized waveforms for improving the range and reliability of RFID systems," in *Proc. IEEE International Conference on RFID, Orlando*, pp. 80–87, 2009.

[148] A. S. Boaventura and N. B. Carvalho, "Maximizing dc power in energy harvesting circuits using multisine excitation," in *Proc. IEEE MTT-S International Microwave Symposium (IMS)*, pp. 1–4, 2011.

[149] A. Collado and A. Georgiadis, "Optimal waveforms for efficient wireless power transmission," *IEEE Microwave and Wireless Components Letters*, vol. 24, no. 5, pp. 354–356, May 2014.

[150] A. Kurs, A. Karalis, R. Moffat, J. D. Joannopoulos, P. Fischer, and M. Soljacic, "Wireless power transfer via strongly coupled magnetic resonators," *Science*, vol. 317, no. 5834, pp. 83–86, 2007.

[151] A. Karalis, J. D. Joannopoulos, and M. Soljacic, "Efficient wireless non-radiative mid-range energy transfer," *Annals of Physics*, vol. 323, no. 1, pp. 34–48, 2008.

[152] H. Haus and W. Huang, "Coupled mode theory," *Proc. IEEE*, vol. 79, no. 10, pp. 1505–1518, October 1991.

[153] M. Kiani and M. Ghovanloo, "The circuit theory behind coupled-mode magnetic resonance-based wireless power transmission," *IEEE Transactions on Circuits and Systems I: Regular Papers*, vol. 59, no. 9, pp. 2065–2074, September 2012.

[154] A. P. Sample, D. A. Meyer, and J. R. Smith, "Analysis, experimental results and range adaptation of magnetically coupled resonators for wireless power transfer," *IEEE Transactions on Industrial Electronics*, vol. 58, no. 2, pp. 544–554, February 2011.

[155] A. K. R. Rakhyani and G. Lazzi, "Multicoil telemetry system for compensation of coil misalignment effect in implantable systems," *IEEE Antennas and Wireless Propagation Letters*, vol. 11, pp. 1675–1678, 2012.

[156] H. Lang, A. Ludwig, and C. D. Sarris, "Convex optimization of wireless power transfer systems with multiple transmitters," *IEEE Transactions on Antennas and Propagation*, vol. 62, no. 9, pp. 4623–4636, September 2014.

[157] Z. Zhang, H. Pang, A. Georgiadis, and C. Cecati, "Wireless power transfer – an overview," *IEEE Transactions on Industrial Electronics*, vol. 66, no. 2, pp. 1044–1058, February 2019.

[158] www.wirelesspowerconsortium.com.

[159] S. S. Mohan, M. del Mar Henshenson, S. P. Boyd, and T. H. Lee, "Simple accurate expressions for planar spiral inductors," *IEEE Journal of Solid-State Circuits*, vol. 34, no. 10, pp. 1419–1424, October 1999.

[160] H. A. Wheeler, "Simple inductance formulas for radio coils," *Proc. Institute of Radio Engineers*, vol. 16, no. 10, pp. 1398–1400, October 1928.

[161] U.-M. Jow and M. Ghovanloo, "Design and optimization of printed spiral coils for efficient transcutaneous inductive power transmission," *IEEE Transactions on Biomedical Engineering*, vol. 1, no. 3, pp. 193–202, September 2007.

[162] T. Beh, M. Kato, T. Imura, and Y. Hori, "Wireless power transfer system via magnetice resonant coupling at fixed resonance frequency-power transfer system based on impedance matching," *World Electr. Veh. J.*, vol. 4, no. 4, pp. 744–753, December 2010.

[163] M. Soma, D. C. Galbraith, and R. L. White, "Radio-frequency coils in implantable devices: misalignment analysis and design procedure," *IEEE Transactions on Biomedical Engineering*, vol. BME-34, no. 4, pp. 276–282, April 1987.

[164] K. Kim, E. Levi, Z. Zabar, and L. Birenbaum, "Mutual inductance of noncoaxial circular coils with constant current density," *IEEE Transactions on Magnetics*, vol. 33, no. 5, pp. 4303–4309, September 1997.

[165] D. M. Pozar, *Microwave Engineering*, 3rd ed. Wiley, September 2010.

[166] S. Shahid, J. Ball, C. Wells, and P. Wen, "Reflection type Q-factor measurement using standard least squares methods," *IET Microwaves, Antennas & Propagation*, vol. 5, no. 4, pp. 426–432, March 2011.

[167] D. Kaifez, "Q-factor measurement with a scalar network analyser," *IEEE Proceedings – Microwaves, Antennas and Propagation*, vol. 142, no. 5, p. 369, October 1995.

[168] F. Grover, *Inductance Calculations: Working Formulas and Tables.* Van Nostrand, 1946.

[169] D. Ahn and S. Hong, "A study on magnetic field repeater in wireless power transfer," *IEEE Transactions on Industrial Electronics*, vol. 60, no. 1, pp. 360–371, January 2013.

[170] Y. Narusue, Y. Kawahara, and T. Asami, "Impedance matching method for any-hop straight wireless power transmission using magnetic resonance," in *Proc. IEEE Radio and Wireless Symposium (RWS), Austin, TX*, pp. 193–195, January 2013.

[171] B. Wang, W. Yerazunis, and K. H. Teo, "Wireless power transfer: Metamaterials and array of coupled resonators," *Proceedings of the IEEE*, vol. 101, no. 6, pp. 1359–1368, June 2013.

[172] J. Park, Y. Tak, Y. Kim, Y. Kim, and S. Nam, "Investigation of adaptive matching methods for near-field wireless power transfer," *IEEE Transactions on Antennas and Propagation*, vol. 59, no. 5, pp. 1769–1773, May 2011.

[173] S. Iguchi, P. Yeon, H. Fuketa, K. Ishida, T. Sakurai, and M. Takamiya, "Zero phase difference capacitance control (ZPDCC) for magnetically resonant wireless power transmission," in *Proc. IEEE Wireless Power Transfer Conference (WPTC)*, pp. 23–28, May 2013.

[174] M. Schuetz, A. Georgiadis, A. Collado, and G. Fischer, "A particle swarm optimizer for tuning a software-defined, highly configurable wireless power transfer platform," in *Proc. IEEE Wireless Power Transfer Conference (WPTC)*, pp. 1–4, 2015.

[175] L. Huang and A. P. Hu, "Defining the mutual coupling of capacitive power transfer for wireless power transfer," *Electronics Letters*, vol. 51, no. 22, pp. 1806–1807, October 2015.

[176] J. Dai and D. C. Ludois, "A survey of wireless power transfer and a critical comparison of inductive and capacitive coupling for small gap applications," *IEEE Transactions on Power Electronics*, vol. 30, no. 11, pp. 6017–6029, November 2015.

[177] ——, "Single active switch power electronics for kilowatt scale capacitive power transfer," *IEEE Journal of Emerging and Selected Topics in Power Electronics*, vol. 3, no. 1, pp. 315–323, March 2015.

[178] N. Shinohara, "Power without wires," *IEEE Microwave Magazine*, vol. 12, no. 7, pp. S64–S73, December 2011.

[179] R. M. Dickinson, "Performance of a high-power 2.388 GHz receiving array in wireless power transmission over 1.54 km," in *Proc. IEEE MTT-S International Microwave Symposium (IMS)*, pp. 139–141, 1976.

[180] T. Ohira, "Power efficiency and optimum load formulas on RF rectifiers featuring flow-angle equations," *IEICE Electronics Express*, vol. 10, no. 11, pp. 1–9, May 2013.

[181] M. Roberg, T. Reveyrand, I. Ramos, E. Falkenstein, and Z. Popovic, "High-efficiency harmonically terminated diode and transistor rectifiers," *IEEE Transactions on Microwave Theory and Techniques*, vol. 60, no. 12, pp. 4043–4052, December 2012.

[182] D. Hamill, "Time reversal duality and the synthesis of a double class E dc–dc converter," in *Proc. 21st Power Electronics Specialist Conference*, pp. 512–521, 1990.

[183] F. Raab, "Class-E, class-C, and class-F power amplifiers based upon a finite number of harmonics," *IEEE Transactions on Microwave Theory and Techniques*, vol. 49, no. 8, pp. 1462–1468, August 2001.

[184] J. O. McSpadden, R. M. Dickinson, L. Fan, and K. Chang, "A novel oscillating rectenna for wireless microwave power transmission," in *Proc. IEEE MTT-S International Microwave Symposium (IMS), Baltimore, MD, USA*, pp. 1161–1164, 1998.

[185] M. D. Prete, A. Costanzo, A. Georgiadis, A. Collado, D. Masotti, and Z. Popovic, "A 2.45 GHz energy-autonomous wireless power relay node," *IEEE Transactions on Microwave Theory and Techniques*, vol. 63, no. 12, pp. 4511–4520, December 2015.

[186] G. Chattopadhyay, H. Manohara, M. M. Mojarradi, et al., "Millimeter-wave wireless power transfer technology for space applications," in *Proc. Asia-Pasific Microwave Conference, Macau*, pp. 1–4, 2008.

[187] A. Collado and A. Georgiadis, "24 GHz substrate integrated waveguide (SIW) rectenna for energy harvesting and wireless power transmission," in *Proc. IEEE MTT-S International Microwave Symposium (IMS), Seattle, WA*, pp. 1–3, 2013.

[188] T. Blackwell, "Recent demonstrations of laser power beaming at DFRC and MSFC," in *AIP Conference Proceedings*, vol. 766, pp. 73–85, 2005.

[189] A. Sample and J. R. Smith, "Experimental results with two wireless power transfer systems," in *Proc. IEEE Radio and Wireless Symposium, San Diego, CA*, pp. 16–18, 2009.

[190] M. Piñuela, P. D. Mitcheson, and S. Lucyszyn, "Ambient RF energy harvesting in urban and semi-urban environments," *IEEE Transactions on Microwave Theory and Techniques*, vol. 61, no. 7, pp. 2715–2726, July 2013.

[191] L. Guenda, E. Santana, A. Collado, K. Niotaki, N. B. Carvalho, and A. Georgiadis, "Electromagnetic energy harvesting–global information database," *Wiley Transactions on Emerging Telecommunications Technologies*, vol. 25, no. 1, pp. 56–63, January 2014.

[192] A. Dolgov, R. Zane, and Z. Popovic, "Power management system for online low power RF energy harvesting optimization," *IEEE Transactions on Circuits and Systems I: Regular Papers*, vol. 57, no. 7, pp. 1802–1811, July 2010.

[193] V. Rizzoli, G. Bichicchi, A. Costanzo, F. Donzelli, and D. Masotti, "CAD of multi-resonator rectenna for micro-power generation," in *Proc. European Microwave Conference, Rome*, 2009, pp. 1684–1687.

[194] Y. H. Suh and K. Chang, "A high efficiency dual frequency rectenna for 2.45 and 5.8-GHz wireless power transmission," *IEEE Transactions on Microwave Theory and Techniques*, vol. 50, no. 7, pp. 1784–1789, July 2002.

[195] C. Gomez, J. Garcia, and A. Mediavilla, "A high efficiency rectenna element using E-pHEMT technology," in *Proc. GAAS, Amsterdam*, pp. 315–318, October 2004.

[196] V. Palazzi, J. Hester, J. Bito, et al., "A novel ultra-lightweight multiband rectenna on paper for RF energy harvesting in the next generation LTE bands," *IEEE Transactions on Microwave Theory and Techniques*, vol. 66, no. 1, pp. 366–379, January 2018.

[197] O. Georgiou, K. Mimis, D. Halls, W. H. Thompson, and D. Gibbins, "How many Wi-Fi APs does it take to light a lightbulb?" *IEEE Access*, vol. 4, pp. 3732–3746, 2016.

[198] U. Olgun, C.-C. Chen, and J. L. Volakis, "Design of an efficient ambient WiFi energy harvesting system," *IET Microwaves, Antennas and Propagation*, vol. 6, no. 11, pp. 1200–1206, August 2012.

[199] K. Mimis, D. Gibbins, S. Dumanli, and G. T. Watkins, "Ambient RF energy harvesting trial in domestic settings," *IET Microwaves, Antennas & Propagation*, vol. 9, no. 5, pp. 454–462, May 2015.

[200] Y. Kawahara, X. Bian, R. Shigeta, R. Vyas, M. M. Tentzeris, and T. Asami, "Power harvesting from microwave oven electromagnetic leakage," in *Proc. 2013 ACM International Joint Conference on Pervasive and Ubiquitous Computing (UbiComp), Zurich*, pp. 373–382, September 2013.

[201] "Monza R6 product brief/datasheet v.5," www.impinj.com, 2017.

[202] A. Boaventura, A. Collado, N. Carvalho, and A. Georgiadis, "Optimum behavior: wireless power transmission system design through behavioral models and efficient synthesis techniques," *IEEE Microwave Magazine*, vol. 14, no. 2, pp. 26–35, March–April 2013.

[203] "Fundamentals of RF and microwave power measurements, Agilent Application Note," Santa Rosa, CA, 2005.

[204] R. G. Harrison and X. L. Polozec, "Nonsquarelaw behavior of diode detectors analyzed by the Ritz–Galerkin method," *IEEE Transactions on Microwave Theory and Techniques*, vol. 42, no. 5, pp. 840–846, May 1994.

[205] J. Akkermans, M. van Beurden, G. Doodeman, and H. Visser, "Analytical models for low-power rectenna design," *IEEE Antennas and Wireless Propagation Letters*, vol. 4, pp. 187–190, 2005.

[206] G. D. Vita and G. Iannaccone, "Design criteria for the RF section of UHF and microwave passive RFID transponders," *IEEE Transactions on Microwave Theory and Techniques*, vol. 53, no. 9, pp. 2978–2990, September 2005.

[207] J.-P. Curty, N. Joehl, F. Krummenacher, C. Dehollain, and M. J. Declercq, "A model for μ-power rectifier analysis and design," *IEEE Transactions on Circuits and Systems I: Regular Papers*, vol. 52, no. 12, pp. 2771–2779, December 2005.

[208] A. M. Cowley and H. O. Sorensen, "Quantitative comparison of solid-state microwave detectors," *IEEE Transactions on Microwave Theory and Techniques*, vol. MTT-14, no. 12, pp. 588–602, December 1966.

[209] F. Bolos, J. Blanco, A. Collado, and A. Georgiadis, "RF energy harvesting from multi-tone and digitally modulated signals," *IEEE Transactions on Microwave Theory and Techniques*, vol. 64, no. 6, pp. 1918–1927, June 2016.

[210] J. Mautz and R. Harrington, "Modal analysis of loaded N-port scatterers," *IEEE Transactions on Antennas and Propagation*, vol. AP-21, no. 2, pp. 188–199, March 1973.

[211] R. Collin, *Antennas and Radiowave Propagation*. McGraw-Hill, 1985.

[212] R. Fano, "Theoretical limitations on the broadband matching of arbitrary impedances," Res. Lab. Electron. MIT, Cambridge, MA, USA, Tech. Rep. 41, January 1948.

[213] K. Niotaki, S. Kim, S. Jeong, A. Collado, A. Georgiadis, and M. M. Tentzeris, "A compact dual-band rectenna using slot-loaded dual band folded dipole antenna," *IEEE Antennas and Wireless Propagation Letters*, vol. 12, pp. 1634–1637, 2013.

[214] B. Pham and A.-V. Pham, "Triple bands antenna and high efficiency rectifier design for RF energy harvesting at 900, 1900 and 2400 MHz," in *Proc. IEEE MTT-S International Microwave Symposium (IMS), Seattle, WA, USA*, pp. 1–3, June 2013.

[215] R. Scheeler, S. Korhummel, and Z. Popovic, "A dual-frequency ultralow-power efficient 0.5-g rectenna," *IEEE Microwave Magazine*, vol. 15, no. 1, pp. 10–114, January–February 2014.

[216] J. Kimionis, A. Collado, M. M. Tentzeris, and A. Georgiadis, "Octave and decade printed UWB rectifiers based on nonuniform transmission lines for energy harvesting," *IEEE Transactions on Microwave Theory and Techniques*, vol. 65, no. 11, pp. 4326–4334, November 2017.

[217] R. Collin, *Foundations for Microwave Engineering*, 2nd ed., ser. IEEE Press Series on Electromagnetic Field Theory. Wiley, 2001.

[218] S. N. Daskalakis, A. Georgiadis, A. Collado, and M. M. Tentzeris, "A UHF rectifier with 100% bandwidth based on a ladder LC impedance matching network," in *Proc. 12th European Microwave Integrated Circuits Conference (EuMIC), Nuremberg*, pp. 411–414, 2017.

[219] H. Yehui, O. Leitermann, D. A. Jackson, J. M. Rivas, and D. J. Perreault, "Resistance compression networks for radio-frequency power conversion," *IEEE Transactions on Power Electronics*, vol. 22, no. 1, pp. 41–53, January 2007.

[220] K. Niotaki, A. Georgiadis, A. Collado, and J. S. Vardakas, "Dual-band resistance compression networks for improved rectifier performance," *IEEE Transactions on Microwave Theory and Techniques*, vol. 62, no. 12, pp. 3512–3521, December 2014.

[221] H. Sun, Z. Zhong, and Y.-X. Guo, "An adaptive reconfigurable rectifier for wireless power transmission," *IEEE Microwave and Wireless Components Letters*, vol. 23, no. 9, pp. 492–494, September 2013.

[222] R. J. Gutmann and J. M. Borrego, "Power combining in an array of microwave power rectifiers," *IEEE Transactions on Microwave Theory and Techniques*, vol. MTT-27, no. 12, pp. 958–968, December 1979.

[223] N. Shinohara and H. Matsumoto, "Dependence of DC output of a rectenna array on the method of interconnection of its array elements," *Elect. Eng. Japan*, vol. 125, no. 1, pp. 9–17, 1998.

[224] B. R. Marshall and G. D. Durgin, "Staggered pattern charge collection: antenna technique to improve RF energy harvesting," in *Proc. IEEE International Conference on RFID, Malaysia*, pp. 30–35, May 2013.

[225] S. N. Daskalakis, A. Georgiadis, G. Goussetis, and M. M. Tentzeris, "A rectifier circuit insensitive to the angle of incidence of incoming waves based on a Wilkinson power combiner," *IEEE Transactions on Microwave Theory and Techniques*, vol. 67, no. 7, pp. 3210–3218, July 2019.

[226] G. A. Vera, A. Georgiadis, A. Collado, and S. Via, "Design of a 2.45 GHz rectenna for electromagnetic (EM) energy scavenging," in *Proc. IEEE Radio and Wireless Symposium (RWS)*, pp. 61–64, 2010.

[227] A. Papoulis and S. U. Pillai, *Probability, Random Variables and Stochastic Processes*, 4th ed. McGraw-Hill, 2002.

[228] "Characterizing digitally modulated signals with CCDF curves," Keysight, USA, Application Note 5968-6875E, January 2000.

[229] J. Proakis, *Digital Communications*, 4th ed. McGraw-Hill, 2000.

[230] C. Valenta, M. M. Morys, and G. D. Durgin, "Theoretical energy-conversion efficiency for energy-harvesting circuits under power-optimized waveform excitation," *IEEE Transactions on Microwave Theory and Techniques*, vol. 63, no. 5, pp. 1758–1767, May 2015.

[231] J. Blanco, F. Bolos, and A. Georgiadis, "Instantaneous power variance and radio frequency to dc conversion efficiency of wireless power transfer systems," *IET Microwaves, Antennas and Propagation*, vol. 10, no. 10, pp. 1065–1070, October 2016.

[232] M. K. Kazimierczuk, *Pulse-Width Modulated DC-DC Converters*, 2nd ed. Wiley, 2016.

[233] R. W. Erickson and D. Maksimovic, *Fundamentals of Power Electronics*, 2nd ed. Kluwer Academic Publishers, 2004.

[234] Y. Huang, N. Shinohara, and T. Mitani, "Theoretical analysis on dc–dc converter for impedance matching of a rectifying circuit in wireless power transfer," in *Proc. IEEE International Symposium on Radio-Frequency Integration Technology (RFIT), Sendai*, pp. 229–231, 2015.

[235] ——, "A constant efficiency of rectifying circuit in an extremely wide load range," *IEEE Transactions on Microwave Theory and Techniques*, vol. 62, no. 4, pp. 986–993, April 2014.

[236] T. Paing, J. Shin, R. Zane, and Z. Popovic, "Resistor emulation approach to low-power RF energy harvesting," *IEEE Transactions on Power Electronics*, vol. 23, no. 3, pp. 1494–1501, May 2008.

[237] L. M. M. Pollak and P. Spies, "Step-up dc–dc converter with coupled inductors for low input voltages," in *Proc. PowerMEMS, Sendai*, pp. 141–148, November 2008.

[238] H. Yu, H. Wu, and Y. Wen, "An ultra-low input voltage power management circuit for indoor micro-light energy harvesting system," in *Proc. IEEE Sensors*, pp. 261–264, 2010.

[239] S. Adami, V. Marian, N. Degrenne, C. Vollaire, B. Allard, and F. Costa, "Self-powered ultra-low power dc–dc converter for RF energy harvesting," in *Proc. IEEE Faible Tension Faible Consommation, Paris*, pp. 1–4, 2012.

[240] E. Koutroulis, K. Kalaitzakis, and N. C. Voulgaris, "Development of a microcontroller-based, photovoltaic maximum power point tracking control system," *IEEE Transactions on Power Electronics*, vol. 16, no. 1, pp. 46–54, January 2001.

[241] L. Roselli, Ed., *Green RFID Systems*. Cambridge University Press, 2014.

[242] M. Winter and R. J. Brodd, "What are batteries, fuel cells and supercapacitors?" *Chemical Reviews*, vol. 104, no. 10, pp. 4245–4269, 2004.

[243] D. Linden and T. B. Reddy, Eds., *Handbook of Batteries*, 3rd ed. McGraw-Hill, 2001.

[244] A. Devices, "LTC1540, nanopower comparator with reference," www.linear .com/LTC1540, 1997.

[245] "Impinj monza R6-P tag chip datasheet, v. 5.0," https://support.impinj .com/hc/en-us/articles/204793258-Monza-R6-P-Product-Brief-Datasheet, 2018.

[246] S. Kim, R. Vyas, J. Bito, et al., "Ambient RF energy-harvesting technologies for self-sustainable standalone wireless sensor platforms," *Proceedings of the IEEE*, vol. 102, no. 11, pp. 1649–1666, November 2014.

[247] A. N. Parks and J. R. Smith, "Sifting through the airwaves: efficient and scalable multiband RF harvesting," in *Proc. IEEE International Conference on RFID*, pp. 74–81, April 2014.

[248] M. T. Penella-Lopez and M. Gasulla-Forner, *Powering Autonomous Sensors: An Integral Approach with Focus on Solar and RF Energy Harvesting.* Springer, 2011.

[249] A. Luque and S. Hegedus, *Handbook of Photovoltaic Science and Engineering*, 2nd ed. Wiley, 2011.

[250] K. Niotaki, F. Giuppi, A. Georgiadis, and A. Collado, "Solar/em energy harvester for autonomous operation of a monitoring sensor platform," *Cambridge Wireless Power Transfer*, vol. 1, no. 1, pp. 1–7, April 2014.

[251] M. Virili, A. Georgiadis, A. Collado, K. Niotaki, P. Mezzanotte, L. Roselli, F. Alimenti, and N. B. Carvalho, "Performance improvement of rectifiers for WPT exploiting thermal energy harvesting," *Cambridge Wireless Power Transfer*, vol. 2, no. 1, pp. 22–31, April 2015.

[252] S. Dunbar and Z. Popovic, "Low power electronics for energy harvesting sensors," *Cambridge Wireless Power Transfer*, vol. 1, no. 1, pp. 35–43, March 2014.

[253] A. P. Sample, D. J. Yeager, P. S. Powledge, A. V. Mamishev, and J. R. Smith, "Design of an RFID-based battery-free programmable sensing platform," *IEEE Transactions on Instrumentation and Measurement*, vol. 57, no. 11, pp. 2608–2615, November 2008.

[254] D. de Donno, L. Catarinucci, and L. Tarriconne, "RAMSES: RFID augmented module for smart environmental sensing," *IEEE Transactions on Instrumentation and Measurement*, vol. 63, no. 7, pp. 1701–1708, July 2014.

[255] G. Marrocco, L. Mattioni, and C. Calabrese, "Multiport sensor RFIDs for wireless passive sensing of objects – basic theory and early results," *IEEE Transactions on Antennas and Propagation*, vol. 56, no. 8, pp. 2691–2702, August 2008.

[256] R. Bhattacharyya, C. Floerkemeier, and S. Sarma, "Low-cost, ubiquitous RFID-tag-antenna-bssed sensing," *Proceedings of the IEEE*, vol. 98, no. 9, pp. 1593–1600, September 2010.

[257] S. Kim, Y. Kawahara, A. Georgiadis, A. Collado, and M. M. Tentzeris, "Low-cost inkjet-printed fully passive RFID tags for calibration-free capacitive/haptic sensor applications," *IEEE Sensors*, vol. 15, no. 6, pp. 3135–3145, June 2015.

[258] D. Pozar, "Scattered and absorbed powers in receiving antennas," *IEEE Antennas and Propagation Magazine*, vol. 46, no. 1, pp. 144–145, February 2004.

[259] K. V. S. Rao, P. V. Nikitin, and S. F. Lam, "Antenna design for UHF RFID tags: a review and a practical application," *IEEE Transactions on Antennas and Propagation*, vol. 53, no. 12, pp. 3870–3876, December 2005.

[260] J. Kimionis, A. Bletsas, and J. N. Sahalos, "Increased range bistatic scatter radio," *IEEE Transactions on Communications*, vol. 62, no. 3, pp. 1091–1104, March 2014.

[261] V. Liu, A. Parks, V. Talla, S. Gollakota, D. Wetherall, and J. R. Smith, "Ambient backscatter: wireless communication out of thin air," in *Proc. ACM SIGCOMM*, pp. 39–50, 2013.

[262] S. N. Daskalakis, J. Kimionis, A. Collado, G. Goussetis, M. M. Tentzeris, and A. Georgiadis, "Ambient backscatterers using FM broadcasting for low cost and low power wireless applications," *IEEE Transactions on Microwave Theory and Techniques*, vol. 65, no. 12, pp. 5251–5262, December 2017.

Index